Edward Money

Essay on the Cultivation and Manufacture of Tea

An Essay for Which the Prize of the Grant Gold Medal and Rs. 300 was Awarded by

the Agricultural and Horticultural Society of India in the year 1872

Edward Money

Essay on the Cultivation and Manufacture of Tea
An Essay for Which the Prize of the Grant Gold Medal and Rs. 300 was Awarded by the Agricultural and Horticultural Society of India in the year 1872

ISBN/EAN: 9783337059354

Printed in Europe, USA, Canada, Australia, Japan

Cover: Foto ©berggeist007 / pixelio.de

More available books at **www.hansebooks.com**

ESSAY

ON THE

CULTIVATION & MANUFACTURE OF TEA,

BY

Lieut.-Colonel EDWARD MONEY,

FOR WHICH THE PRIZE OF THE GRANT GOLD MEDAL AND
Rs. 300 WAS AWARDED BY THE AGRICULTURAL
AND HORTICULTURAL SOCIETY OF INDIA
IN THE YEAR 1872.

SECOND EDITION,
Corrected and much enlarged.

PRICE RS. 4.

Calcutta:
WYMAN & Co., PUBLISHERS,
5, COUNCIL HOUSE STREET, & 10, HARE STREET.

1874.

TABLE OF CONTENTS.

	Chapter.
Past and present financial prospects of Tea	I.
Labor, local and imported	II.
Tea districts and their comparative advantages. Climate, soil, &c., in each	III.
Soil	IV.
Nature of jungle	V.
Water and sanitation	VI.
Lay of land	VII.
Laying out a garden	VIII.
Varieties of the Tea-plant	IX.
Tea seed	X.
Comparison between sowing in nurseries and in situ	XI.
Sowing seed in situ, *id. est.* at stake	XII.
Nurseries	XIII.
Manure	XIV.
Distances apart to plant Tea-bushes	XV.
Making a garden	XVI.
Transplanting	XVII.
Cultivation of made gardens	XVIII.
Pruning	XIX.
White-ants, crickets, and blight	XX.
Filling up vacancies	XXI.
Flushing and number of flushes	XXII.
Leaf-picking	XXIII.
Manufacture. Mechanical contrivances	XXIV.
Sifting and sorting	XXV.
Boxes, packing	XXVI.
Management, accounts, forms	XXVII.
Cost of manufacture, packing, transport, &c.	XXVIII.
Cost of making a 300 acre Tea garden	XXIX.
How much profit Tea can give	XXX.
The past, present, and future of Indian Tea	XXXI.

PREFACE TO FIRST EDITION.

THE following Essay was written with, *firstly*, the object of competing for the Gold Medal and the Money Prize offered by the Agricultural and Horticultural Society of India for the best treatise on the cultivation and manufacture of Tea; and, *secondly*, with the view of arranging the hundreds of notes on these subjects, which, in the course of eleven years, I had collected.

During all these years I have been a Tea Planter, making first for myself and others a garden in the Himalayas, and for the last six years doing the same thing for myself in the Chittagong district.

Whenever I have visited other plantations, (and I have seen a great number in many districts), I have brought away notes of all I saw. Up to the last, at every such visit, I have learnt *something*. If rarely nothing to follow, something at least to avoid. I have now tested all and everything connected with the cultivation and manufacture of Tea by my own experience, and I can only hope that what I have written will be found useful to an industry, destined, yet I believe, in spite of the late panic, the natural result of wild speculation, to play an important part in India.

I have endeavored to adapt this Essay to the wants of a beginner, as there are many of that class now, and may yet be more in days to come, who must feel, as I often have, the want of a really practical work on Tea.

To those who have Tea properties in unlikely climates and unlikely sites, I would say two words. No view I have taken of the advantages of different localities, *can* in any way affect the results of enterprises already entered upon. But if the note of warning, sounded in the following pages, checks further losses in Tea, already so vast, while it fosters the cultivation on remunerative sites, I shall not have written in vain.

EDWARD MONEY.

SUNGOO RIVER PLANTATION;
   CHITTAGONG,
   November 1870.

PREFACE TO SECOND EDITION.

THREE years' further experience, and visiting two Tea districts I had not seen before, has enabled me to amend whatever was faulty in the 1st Edition. The whole has been revised, and much new matter is added throughout. A new Chapter at the end on the past, the present, and the future of Indian Tea will, it is hoped, be found interesting. An Index (a great want in the 1st Edition) is added, so that all information on any point can be at once found. The manufacture of Green Tea, which I was ignorant of when I last wrote, is given, and the advisability of that manufacture is discussed.

In its present form I hope and believe this little work will be found useful and interesting to all connected with Tea.

EDWARD MONEY.

DARJEELING,
   May 1874.

ADDENDA.

Since this second and enlarged edition went to Press, some facts and information have been met with by me, and I give them here.

DISSOLVED PERUVIAN GUANO,

This is a manure highly spoken of in England, and for which Messrs. Ede and Hobson are the Agents in Calcutta. It appears to be a highly concentrated article, and it may be that it will prove very beneficial to the Tea-plant. It should certainly be tried by planters; I am trying it myself, but regret that as I have only just applied it I cannot speak as to the results.

The following is the description given by the manufacturers:—

DISSOLVED PERUVIAN GUANO,

PREPARED BY OHLENDORFF & Co.,

LONDON, ANTWERP, HAMBURG, AND EMMERICH-ON-RHINE.

GUARANTEED

TO CONTAIN:

Nitrogen equal to 10 per cent. of non-volatile ammonia,

20 per cent. of soluble } *Guano Phosphate.*
4 per cent. of insoluble }

Our Dissolved Peruvian Guano has, since we first introduced it ten years ago, met with such unqualified success on the Continent, and more specially in Germany, that we are induced to come forward with it before the British public, and

ii ADDENDA.

with this purpose in view we have errected works at Plaistow, near Victoria Docks, London.

General character of the Dissolved Peruvian Guano.

This manure is prepared from *genuine Peruvian Government Guano*, treated with sufficient sulphuric acid, to fix the ammonia, and to convert the insoluble fertilizing constituents of the raw Guano into readily available soluble compounds. It supplies the farmer with the means of deriving the greatest economical advantage from the use of Guano.

It is offered for sale as "*Dissolved Peruvian Guano*," in a nicely-prepared and dry condition, and of an uniform strength, guaranteed as above.

Peculiar merits of the Dissolved Peruvian Guano.

The following are some of the merits of the Dissolved Peruvian Guano which have been fully recognised by all persons, who have had fair opportunities to become practically acquainted with this valuable and concentrated manure:—

1. It contains a very high percentage of ammonia—the most valuable fertilizing constituent of most natural and artificial manures.

2. All the ammonia occurs in Dissolved Peruvian Guano in a fixed or non-volatile, and at the same time readily soluble and available condition. Dissolved Guano for this reason loses nothing whatever by exposure to the air, nor even by exposure to a burning sun. If kept in a dry place its fertilizing properties remain unimpaired for years in all climates.

3. It is as rich in soluble phosphate as good superphosphates and most dissolved bone manures, and thus combines the fertilizing properties of phosphatic manures with those of raw Peruvian Guano, and of other ammoniacal manures.

4. In addition to a high percentage of non-volatile ammonia and of a large percentage of soluble Guano phosphate, Dissolved Peruvian Guano contains a useful amount of *insoluble Guano phosphates* to support the healthy growth of cultivated plants at the later stages of their development; and it further contains soluble silica, a large proportion of nitrogenous organic

matters, and a considerable amount of soluble salts, of potash, soda, and magnesia. It thus embodies all the essential organic and mineral elements of plant-food which are required to encourage the healthy and luxuriant growth of our crops.

5. Inasmuch as Dissolved Peruvian Guano contains all the elements of nutrition to favor the abundant yield of all kinds of agricultural produce, it is more useful as a fertilizer for general agricultural purposes than special artificial manures, whose efficiency depends entirely upon the amount of soluble phosphate, or of potash, or of ammonia, or of any one particular fertilizing constituent, for which many artificial manures are specially recommended.

6. It is prepared and sold of uniform strength guaranteed to contain—

Nitrogen equal to 10 per cent. of non-volatile ammonia,

    20 per cent. of soluble   } *Guano phosphate.*
    4 per cent. of insoluble

In respect of this guarantee we hold ourselves bound by the result of Analyses of the Chemists to the three great National Agricultural Societies, *i.e.*—

Dr. AUGUSTUS VOELCKER, *Consulting Chemist to the Royal Agricultural Society of England, London.*

Prof. THOS. ANDERSON, *Chemist of the Highland and Agricultural Society of Scotland, Glasgow.*

Dr. JAS APJOHN, *Chemist of the Royal Agricultural Society of Ireland, Dublin.*

7. Dissolved Peruvain Guano is a dry and finely-prepared manure, which requires no previous breaking up, sifting, and similar manipulation, before it can be applied beneficially to the crop for which it is intended.

8. It may be sown broadcast and left on the surface of the land exposed to the air and sun, without losing any of its fertilizing constituents, and need not be ploughed in or harrowed in at once, but the operation of mixing it with the surface soil by a light harrowing in may be deferred to a period most convenient to the farmer.

iv ADDENDA.

9. For the convenience of dealers the Dissolved Peruvian Guano is made up in bags of uniform weight of 160 lbs. gross each.
For further particulars we crave reference to our pamphlet, containing certificates from the most prominent English and Foreign Agricultural Chemists, with their opinions of our Dissolved Peruvian Guano; this pamphlet will be sent free on application.

OHLENDORFF & Co.
110, FENCHURCH STREET, E.C.,
LONDON, *October*, 1873.

The Imports and Deliveries of Indian Teas into Great Britain have been as following during the last four years :—

	1873.	1872.	1871.	1870.
Imports ...	18,367,000	16,942,000	15,457,000	13,148,168
Deliveries ...	18,187,000	16,276,000	13,706,000	13,472,800

The above shows that the deliveries have kept pace with the imports, and shows, moreover, that the increase in the imports during that time is under two millions of pounds yearly. The estimate of produce for 1874 (see page 176) of "not far short of twenty millions of pounds" will now, I think, be scarcely attained, for the last weather accounts from Assam are far from good. If so, the increase this year (1874) will be again considerably under two millions.

The following regarding Tea cultivation in Java is interesting :—

MESSRS. THOMAS WATSON & Co. report that the quantity of Tea grown in Java and imported into Holland has largely increased. The imports of Java Tea into Holland amounted to 2,541,000lbs. in 1869, and 6,023,500lbs. in 1873.

The following from the "*Indian Economist*," regarding Indian Teas in general, and Neilgherry Teas in particular, is not out of place here. At the same time I do not agree with the writer, for I believe that in the strength and pungency of Indian Teas consists their value :—

INDIAN TEA.

"That the Teas of India have at length come to be fully appreciated in England may be taken, we presume, as an admitted fact ; and it is of importance that planters should direct their attention to modifying their methods of manufacture so as to suit the public taste, and, if possible, turn out an article free from the objections still advanced against the Indian leaf as a daily beverage. There are, we know, those who argue that enough has been done, and that consumers will acquire a taste for the produce of our gardens in time; but we have daily evidence that in the most trivial matters there is no greater tyrant than the public. It behoves those then who cater for this tyrant to consult its taste and satisfy its demands, however exacting and capricious they may be. The remarks we are about to make are based on experiments and enquiries extending over some years in this country and in England, and we leave those engaged in the enterprise to estimate their value. All Teas grown in the plains of India are known to the trade in London under the general name of Assam, and are chiefly used for mixing, seldom reaching the consumer in a pure state. When they do, the objections raised are, that the leaf is too pungent and rough for most palates ; and purchasers are in the habit of mixing it with China to tone down those astringent qualities. In other words, it wants the delicacy of flavour which is the chief characteristic of the Chinese leaf, meaning of course

ADDENDA

that vended by respectable houses, not the abominable trash that formed part of the cargoes of the *Lalla Rookh*, and *Sarpedon* containing, according to Dr. Letheby's Analysis, "40 to 45 per cent. of iron filings and 19 per cent. of silica." Nor is this lack of delicacy of flavour to be lightly regarded, for the efforts of our manufacturers have been directed unwittily and indirectly to foster the peculiarity, as the test of Indian Tea has hitherto been its strength and pungency, to fit it for salting weak, thin, inferior sorts of China. This is what the dealers have demanded, and what, consequently, brokers in their turn have insisted on, with the result that the outturn of our Assam and Cachar plantations, is now, if anything, too powerful to suit public taste. Whether means of manipulation may be hit upon by which aroma can be retained without sacrificing strength, we leave those most interested to determine; but it is worthy of note that this objection to strength and roughness is almost confined to women, the sterner sex preferring Assam unmixed, while the working classes of both sexes are unanimous in favour of the unadulterated Indian article. Experiments were further tried by substituting Neilgherry Tea, and after a short interval the verdict of the majority was in its favour. We need now only point out the difference in the manufacture between the two Teas, leaving others to decide questions regarding the bearing of climate or altitude. Up to the time of finishing rolling the manipulation of the leaf is identical, care being taken to retain the juice; but that made on the hills instead of being almost immediately placed over *choolas* was spread out thinly on tables all night, in a temperature of 54 deg., sustaining consequent loss of strength by evaporation, but developing an aroma that established it at once in favour. So successful has this Neilgherry Tea been at home, that offers are now received by plantation proprietors for their produce at half a crown per lb. free on board, in Madras. This would seem to indicate that the aroma is generated by the action of cold upon the damp leaf while in a state of "suspended fermentation;" for, previous to experimenting with consumers, the samples were submitted to

Mincing Lane brokers and pronounced sound, in corroboration of which opinion the bulk from which they were taken sold at auction for 2s. 2½d., so that fermentation, i.e. (Sourness) had been carefully avoided. We know that the climate of Assam and temperature of the Tea-houses render the keeping of rolled leaf even for an hour fatal to soundness; but should the development of this aroma be really due to "suspension of fermentation," is it not worth while adopting some contrivance for cooling down a chamber set aside for the purpose of spreading out the rolled leaf to the temperature required?

"The question whether delicacy is due to altitude alone and not to manufacture might be ascertained by experiment. Let a quantity of green leaf be sent *down* from one of the Neilgherry gardens, and worked up in the plains at the foot of the hills, and an equal quantity sent *up* from one of the Assam gardens, say to Shillong, and manufactured on the Neilgherry principles there, and the result then compared. This experiment would cost little and determine a not unimportant question: for all engaged in Tea are interested in using their best endeavors to fit it for public consumption, and to guard it against China in any shape or form whatever."

That "delicacy of flavour," and "want of strength" with it, *is* due to altitude has long ago been admitted, and any experiments on that head would, I think, be quite unnecessary. The experiments as to manufacture on the Neilgherris are interesting, and should be further looked into.

The following extract from the "*Darjeeling News*" of May 30th, 1874, describes (and describes well) "Jackson's Rolling Machine," the last rolling machine invented:—

"We have had the advantage of seeing one of Jackson's Tea-Rolling Machines at work during the week, and as we know that several have been ordered, and are actually on their way out for different estates in this district, a short description of the way in which the machine gets through its work may be interesting to some of our readers, who, like ourselves, take an interest in the application of machinery to Tea manufacture. The rolling table of this machine is, like that of other machines, familiar to tea planters, supported by counterpoises, but unlike the others it slides forwards and downwards when it is necessary to take the rolled leaf away, thus forming an inclined plane from which the leaf can be swept into the trays. The chief peculiarity of the table is, that in the centre there is a recess of some few inches in depth; this is an improvement on the cross pieces fixed on Kinman's table. Motion is given from the fly-wheel to the upper portion of the machine by one crank behind having a direct front-and-back action, while another crank is connected with the side of the machine by a ball and socket joint. The upper part of the machine by which the rolling is done is a square box, having a piece of corrugated iron, shaped like the upper shell of a turtle, connected to the inner and lower sides; this box is suspended from a rectangular iron frame by four iron uprights, each upright having two ball and socket joints. These are the principal points which would strike an observer when the machine is at rest. The motions of the upper portion of the machine seem to be a combination of a circle and an oval, though by substituting a different cog wheel it is capable of being made to go through no less than twenty-seven different motions. The machine does not appear to scatter the leaf, when it begins rolling, nearly so much as the others do; it does not crush the leaf in the same way even if, as is often the case, the latter be hard; nor does it require so many men to tend it. It is capable of rolling 120lbs. of soft withered leaf in fifteen minutes, but if the leaf is hard it requires something over twenty minutes to do the same amount of work. The cost of the machine is about Rs. 800, and it certainly appears to do all that is claimed for it. One of its peculiarities is the very great smoothness with which it does its work, and the very complete manner in which it rolls even hard leaf, no auxilliary hand rolling being necessary. We are convinced that this machine is the best yet introduced into the district, though at the same time we consider it very far from being perfection. Machinery is certain to be adopted more and more every year in this district, though many planters still think that rolling cannot be done nearly as well by machinery as by hand. This may perhaps be true at present, but there is no reason, where the attention of practical mechanics directed to the subject, why the same opinion should continue to be well founded much longer. A prejudice has always existed against the introduction of machinery into any branch of industry, but it has always been overcome in the end. It will probably be found that Jackson's Rolling Machine saves the wages of a very large number of highly paid men."

The machine above described is the first imported into the Darjeeling district. I have seen and studied it closely. It is certainly by far the best "rolling machine" yet invented.

A new sorting machine is also advertized by the same maker, but I have not seen it.

I have at last completed experiments with a view to do away with the use of charcoal in Tea manufacture, and I think with success.

The "Furnace Teas," for so I purpose naming them, have in most cases been pronounced by the Calcutta brokers to be superior to similar samples of the same day's leaf, made in the usual way over charcoal.

Nothing but the heat generated by *any* fuel placed in furnaces sunk under ground outside the Tea-house, is used. No motive power of any kind is employed. The apparatus is very simple. It is cheap to erect and very durable in character.

As the apparatus, with which the Teas up to the present time have been made, is a rude and imperfect one, having dis-

ADDENDA.

advantages which must tell more or less on the excellence of the Teas so manufactured, and as, even with these disadvantages, the Teas are pronounced by the brokers *at least* equal to charcoal-dried Teas, it is not too much to hope that with a perfect apparatus (one of which will be erected immediately) Teas will be improved in value by this new invention. The following will be shortly the advantages of this new process, even supposing the Teas are no better:—

1. *Economy.*—This will possibly be even greater than what is set out in the extract of the local paper below; for the fact that the Tea is never placed over charcoal until the whole is ignited and has become "live charcoal," is not there recognized, much of the caloric thus escapes.

2. Cleanliness and absence of charcoal dust.
3. Absence of the objectionable fumes of charcoal.
4. Immunity from fire in Tea-houses.
5. Greater speed in the firing process, and the saving of all the labour employed to make charcoal.
6. Reduced temperature in Tea-houses.

If all the above advantages are, as I expect they will be, attained the life of a Tea planter will be more pleasant than hitherto.

The following is the opinion of the new process expressed by the "*Darjeeling News*" of 1st August.

"It has long been a question, which all planters were desirous to solve, if the fumes of charcoal were necessary to make Tea, that is to say if any chemical action was produced on the tea by the said fumes, and if not, whether it would not be possible to do the firing in some other and far cheaper way.

"The question has, we believe, been solved by Colonel Edward Money, and if so, for the invention is quite a new one, a boon of great magnitude will have been conferred on the Tea interest in India. We congratulate this district as being the birth-place of the improvement.

"The apparatus at present in use at Soom, and which we have seen working, is a rough and crude one made on the spot. This, and the more perfect plans from which larger and better ones are to be made, are readily shown by Colonel Money to

ADDENDA.   xi

any one visiting Soom; but until the invention is patented it is not well to describe it in print. Suffice if we say that the invention is a remarkably simple one—cheap to erect—durable in its character, and the working thereof unattended with any expense whatever, beyond the cost of the fuel (which may be of any kind) and which of course will be many times less than charcoal.

"If true, as we hear, that it takes 3¼ maunds of wood generally to make one maund of charcoal, and if also true, as Colonel Money suggests, that the caloric in one maund of wood equals the caloric in two maunds of charcoal, it then follows that each maund of wood, put into Colonel Money's furnace, equals seven maunds of wood used to make charcoal!

"Of course the above are more or less random figures, but they suffice to show that the saving in fuel will be very great—a boon of course to planters, but a boon also to the Forest Department, and to India.

"We knew of the invention some time back, but we forbore to notice it until the broker's reports on the Tea so made had been received. We have now seen these. Samples of "charcoal" and "furnace" Tea were sent down, made from the same leaf, the same day, and manufactured in one up to the "firing" process. Two brokers give the higher value to the furnace Tea, one to the charcoal kind—but the difference is small.

"We believe, as one of our most experienced planters, who has tasted the Teas, been to Soom, and seen the broker's reports, says: that 'the Tea dried by the furnace apparatus will be *at least* equal to that prepared over charcoal.'

"As Colonel Money is already known as an authority in Tea, and as he has stated to us his belief that "charcoal days" for Tea are now at an end, we await with confidence the ultimate success of his invention which even if it makes no better Tea' will certainly make it far cheaper, while the dirt from charcoal dust will be done away with, the temperature of Tea-houses much reduced, and the deleterious fumes of charcoal, so very objectionable from a sanitary point of view in Tea manufacture, will be known no more."

xii   ADDENDA.

Again 29th August, a month later, the "*Darjeeling News*" further remarks:—

"We alluded recently to Colonel Money's very ingenious plan for drying Tea without charcoal. Since then his apparatus has been in full work at Soom, and has been inspected by numbers of the Darjeeling planters, one and all of whom have, we understand, reported most favorably on its working. Samples of Tea manufactured have been from time to time sent to Calcutta brokers for their opinion, and reports have been received from fifteen, of whom seven are in favor of Tea made by the old charcoal process, seven are in favor of the new furnace process, and one reports that the Tea made by each process is exactly the same.

"Colonel Money has now obtained permission from the Soom Board of Directors to erect his improved furnace, which will be in working order by the end of September, and the whole October crop of Soom Tea will be fired by the new furnace.

"Colonel Money has applied for a patent, and as soon as this is granted we hope to give our readers a description of the apparatus. For obvious reasons it would not be advisible to do so before then. We may mention here that one of the most intelligent and practical planters in the district has ordered one of Colonel Money's flues for his private garden.

"Of the commercial success of Colonel Money's apparatus we have no doubt whatever, and we trust that Colonel Money will reap a handsome profit from his very ingenious invention, which will be an undoubted boon not only to this district but to all the Tea-producing districts of India.

"One point which has struck us as good in Colonel Money's apparatus is that the temperature of the Tea-house is considerably lowered during the firing process as compared with the open *chulas*, and that there is no free carbonic acid gas allowed to escape into the Tea-house, so that those very unpleasant symptoms of slow poisoning which often show themselves in planters and Tea-makers will be unknown in future. At our suggestion Colonel Money has decided to keep a register of the maximum temperature of the Tea-house, whilst the open *chulas* continue in use, and to compare it with the temperature when the new apparatus has superseded them, also to test for free carbonic acid gas in the air with each process.

"We are convinced that when the figures are available our readers will be rather astonished at the difference from a sanitary point of view.

"On the whole, we think that Colonel Money's invention is by far the most important application of *common sense* and scientific knowledge to Tea manufacture that we have seen yet, and we are almost certain that his apparatus will before long be adopted throughout the Indian Tea districts."

PRIZE ESSAY

ON THE

CULTIVATION & MANUFACTURE OF TEA

IN

INDIA.

Premium: Three Hundred Rupees and the Grant Gold Medal.

CHAPTER I.

PAST AND PRESENT FINANCIAL PROSPECTS OF TEA.

Will Tea pay? Certainly on a suitable site, and in a good Tea climate; equally certainly *not* in a bad locality with other drawbacks.

Why then has Tea only paid during the last few years (?) Simply because nothing will pay, which is embarked on without the requisite knowledge, and this was pre-eminently the case with Tea.

NOTHING was known of Tea formerly when everybody rushed into it; not much is known even now. Still with those drawbacks and many others the enterprise has survived, and it is very certain the day will never come that Tea cultivation will cease in India.

I believe there is nothing will pay better than Tea if embarked on with the necessary knowledge in suitable places, but failing either of these success must not be hoped for.

It was madness to expect aught but ruin, under the conditions which the cultivation was entered on in the Tea-fever days. People who had failed in everything else were thought quite competent to make plantations. 'Tis true Tea was so entirely a new thing at that time, but few could be found who had any knowledge of it. Still had managers with some practice in agriculture been chosen the end would not have been so disastrous. But any one—literally any one—was taken, and Tea planters in those days were a strange medley of retired or cashiered army and navy officers, medical men, engineers, veterinary surgeons, steamer captains, chemists, shop-keepers of all kinds, stable-keepers, used-up policemen, clerks, and goodness knows what besides!?

Is it strange the enterprize failed in their hands? Would it not have been much stranger if it had not?

This was only one of the many necessities for failure. I call them "necessities" as they appear to have been so industriously sought after, in some cases. I must detail them shortly, for to expatiate on them would fill a book.

No garden should exceed 300 acres under Tea. If highly cultivated one of even half that size will pay enormously, far better than 800 acres of Tea with low cultivation. Add, say 400 acres for charcoal, &c., making 700 acres the outside area that can be required, and the outside that should ever have been purchased for any one estate. Instead of this, individuals and Companies rushing into Tea bought tracts of five, ten, fifteen, and twenty thousand acres. The idea was that though it might not be all cultivated by taking up so large an area all the local labour where there was any would be secured. Often, however, these large tracts were purchased where local labour there was none, and what the object there was is a mystery. I conceive, however, there was a hazy idea that if 500 acres paid well, 1,000 would pay double, and that eventually even two or three thousand acres would be put under Tea and make the fortunate possessor a *millionaire*. In short, there were no bounds—in fancy— to the size a garden might be made, and thus loss No. 2

PROSPECTS OF TEA. 3

took place, when absurdly large areas were bought of the Government and large areas cultivated.

The only fair rules for the sale of waste lands were those of Lord Canning, which the Secretary of State at home, who could know nothing of the subject, chose to modify and upset. Instead of Rs. 2-8 per acre for all waste lands (by no means a low price, when the cost of land in the Colonies is considered) and that the applicant for the land (who had, perhaps, spent months seeking for it) should have it; the illiberal and unjust method of putting the land up to auction with an upset price of Rs. 2-8 was adopted, the unfortunate seeker, finder, and applicant, through whose labour the land had been found, having no advantage over any other bidder. The best, at least the most successful plan in those days, though as unfair and illiberal as the Government action, was to wait till some one, who was supposed to know what good Tea land was, applied for a piece, and then bid half an anna more than he did, and thus secure it. It *paid* much better than hunting about for one-self, and it was kind and considerate on the part of Government to devise such a plan!

In those fever days, with the auction system, lands almost always sold far above their value. The most absurd prices, Rs. 10, and upwards per acre, were sometimes paid for wild jungle lands. Tracts, which natives could have, and in some cases did lease from Government for inconceivably small sums, representing say at 30 years' purchase, 4 annas per acre, were put up for auction with a limit of Rs. 2-8, and sold perhaps at Rs. 8 or 10 per acre. Had the Government given lands *gratis* to Tea cultivators the policy would have been a wise one. To do what they did was scarcely acting up to their professed wish "to develop the resources of the country."

Since the above was written new rules have been published for the sale of waste lands. The objectionable auction system is continued, and the upset price is much enhanced, as follows:—

B

Schedule of Rates of Upset Prices.

	Upset price per acre.
Districts of the Assam Division ...	~~Rs~~ 8
Districts of Cachar and Sylhet ...	,, 8
Districts of the Chittagong Division ...	,, 6
Districts of the Chota Nagpore Division ...	,, 5
The Soonderbuns	,, 5
All other Districts	,, 10

It is not likely that Government will sell much land at such exorbitant rates.

Security of title, it is generally thought, is one of the advantages of buying land from the State; but I grieve to state my experience is that the reverse is the case, and will so remain until the following is done:—

First.—The Government should learn *what is and what is not theirs to sell.* Such an absurdity then, as Government ascertaining years after the auction that they had sold lands they had no right to sell, could not be.

Secondly.—That before land is sold it be properly surveyed and demarcated; and what might so easily have been done, and which alone would have compensated for much of bad procedure in other respects, that the simple and obvious plan, before the sale, of sending a European official to show the neighbouring villagers and intending purchasers, the boundaries of the land to be sold be resorted to.

This last simple expedient would have saved some grantees years of litigation, and many a hard thought of the said grantees against the Government. It would naturally occur to any one at all conversant with the subject; but alas in India this is often not the condition under which laws are made.

But there is another difficulty at the back of all this. Though the Waste Land Rules enact that the Government, and not the grantee, shall be the defendant in any claim for

land within a lot sold, practically the said enactment in no
way saves grantees from litigation. Claimants for land
always plead that it is *not* within the boundaries of the land
sold, and *ergo* the grantee is made the defendant to prove
that it *is*. The villagers never having been shown the
boundaries by any Government official, (for it is not enacted
in the Waste Land Rules) the question whether the land
claimed is within or without the boundaries is an open one,
not always easily decided, and the suit runs its course.

I even know of cases where, though survey has been
charged for at the exorbitant rate of four annas an acre, the
outer boundaries of the lot have never been surveyed at all,
but merely copied from old Collectorate maps, which showed
the boundaries between the zemindaree and waste lands.* Is
it strange then if buying lands from Government is often
buying litigation, worry, loss of time, and money?

In many countries, for example Prussia, (there I know it is
so, for I have tested it again and again) there are official
records, which can and do show to whom any land in question
belongs. This may scarcely be practicable in India, but surely
the question of title being, as it is, in a far worse state in
India than in most countries, any change would be for the
better. Anyhow the present mode the Government adopts
in selling lands is a grievous wrong to the purchasers. Words
cannot describe the worry and loss some have suffered thereby,
and it might all be so easily avoided.

I have above detailed two of the drawbacks Tea had to contend with in its infancy; the absurdly high price paid for land
was the third. The title-difficulty is as bad to-day as formerly.

Again, Companies and proprietors of gardens wishing to
have large areas under cultivation gave their managers
simple orders to extend not judiciously, but in any case.
What was the result? Gardens might be seen in those days

---

* I need scarcely observe it is impossible to define lands from maps alone without the field-book.

with 200 acres of so-called cultivation, but with 60 or even 70 per cent. vacancies, in which the greater part of the labour available was employed in clearing jungle for 100 acres further extension in the following spring. I have seen no garden in Assam or Cachar with less than 20 per cent. vacancies, many with far more, and yet most of them were extending. I do not believe now any garden in all India exists with less than 12 per cent. vacancies, but a plantation as full as this did not exist formerly.

As the expenditure on a garden is in direct proportion to the area cultivated, and the yield of Tea likewise in direct proportion to the number of plants, it follows the course adopted was the one exactly calculated to entail the greatest expenditure for the smallest yield. This unnecessary, this wilful extension, was the fourth and a very serious drawback.

Under this head the fourth drawback may also be included the fact that the weeds in all plantations were ahead of the labour, that is to say, that gardens were not kept clean. This is more or less even the case to-day; it was the invariable rule then. The consequence was two-fold—*first*, a small yield of Tea; *secondly*, an increased expenditure; for it is a fact that the land 50 men can keep always clean, if the weeds are never allowed to grow to maturity and seed, will take nearer 100 if the weeds once get ahead. The results too differ widely: in the first case the soil is always clear; in the second clear only at intervals. The first, as observed, can be accomplished with 50, the latter will take nearly double the men.

The fifth drawback I shall advert to again later, *viz.*, the selection of sloping land, often the steepest that could be found on which to plant Tea. The great mischief thus entailed will be fully described elsewhere. It was the fifth, and not the least, antagonistic point to success.

Number six was the difficulty in the transport of seed to any new locality, for nine times out of ten a large proportion failed; and again the enormous cost of Tea seed in those days, Rs. 200 a maund (Rs. 500 at least, deducting what failed, was its real price). This item of seed alone entailed an enormous outlay, and was the sixth difficulty Tea cultivation had to contend with. It was, however, a source of great profit to the old plantations, and principally accounts for the large dividends paid for years by the Assam Company.

Again, many managers at that time had no experience to guide them in the manufacture of Tea, each made it his own way, and often turned out most worthless stuff. There is great ignorance on the subject at the present time, but those who know *least* to-day, know *more* than the best informed in the Tea-fever period. Indian Tea was a new thing then, the supply was small, and it fetched comparatively much higher prices than it does now. Still much of it was so bad that the average price all round was low.

Tea manufacture, moreover, as generally practised then, was a much more elaborate and expensive process than it is now.

This will be explained further on under the head of Tea manufacture; I merely now state the fact in support of the assertion that the bad Tea made in those days, and the expensive way it was done, was the seventh hindrance to successful Tea cultivation.

Often in those days was a small garden made of 30 or 40 acres, and sold to a Company as 150 or 200 acres! I am not joking. It was done over and over again. The price paid, moreover, was quite out of proportion to even the supposed area. Two or three lakhs of rupees (£20,000 or £30,000) have been often paid for such gardens, when not more than two years old, and 40 per cent. of the existing area, vacancies. The original cultivators "retired," and the Company carried on. With such a drag upon them (apart from all the other drawbacks enumerated) could success be even hoped for? Certainly not.

I could tell of more difficulties the cultivation had to contend with at the outset, but I have said enough to show, as I remarked, "that it was not strange Tea enterprize failed, inasmuch as it would have been much stranger if it had not."

Do any of the difficulties enumerated exist now? And may a person embarking in Tea to-day hope, with reasonable hope, for success? Yes, certainly, I think as regards the latter—the former let us look into.

People who understand more or less of Tea are plentiful, and a good manager, who knows Tea cultivation and Tea manufacture well, may be found. It will scarcely pay to buy land of the Government at the present high rates, but many people hold large tracts in good Tea localities, and would readily sell.

There is plenty of flat land to be got, so no evil from slopes need be incurred.

Tea seed is plentiful though dear.

The manufacture of Tea (though still progessing) is simple, economical, and more or less known. Anyhow a beginner now will commence where others have left off.

Of course to buy a made garden cheap is better than to make one; but the result in this case is of course no criterion of what profit may be expected from Tea cultivation.

As many of the items to be calculated under the heads of cultivation, manufacture, and receipts will be better understood, after details on these subjects are gone into, I shall reserve the consideration of "how much profit Tea can give" to the end of this treatise.

CHAPTER II.

LABOR, LOCAL AND IMPORTED.

WHEN the very large amount of labour required to carry on a plantation is considered, it is evident that facilities for it are a *sine quâ non* to success. Assam and Cachar, the two largest Tea districts, are very thinly populated, and almost entirely dependent on imported labor. The expense of this is great, and it is the one, and consequently a great drawback to those provinces. The only district I know of with a good Tea climate and abundance of local labor is Chittagong. Several other places have a good supply of local labor, but then their climates are not very suitable.

Each coolie imported costs Rs. 30 and upwards (it used to be much more) ere he arrives on the garden and does any work. After arrival he has to be housed; to be cared for and *physiqued* when sick; to be paid when ill as when working; to have work found for him or paid to sit idle when there is no work; and in addition to all this every death, every desertion, is a loss to the garden of the whole sum expended in bringing the man or woman. Contrast this with the advantages of local labor. In many cases no expense for buildings is necessary, as the laborers come daily to work from adjacent villages, and in such cases no expense is entailed by sick men, for these simply remain at home. There is no loss by death or desertions. When no work is required on the garden labor is simply not employed. All this makes local labor, even where the rate of wages is high, very much cheaper than imported.

The action of Government in the matter of imported labor has much increased the difficulties, and expense necessarily attendant on it. It is a vexed and a very long question which I care not to enter into minutely, for it has been discussed already *ad nauseam*; still I must put on record my opinion after looking very closely into it that the Government has not acted wisely, inasmuch as any State interference, in the relations of employer and employed, (outside the protection which the existing laws give) is a radical mistake. As for the law passed on the subject to the effect that a coolie who has worked out his agreement and voluntarily enters into a new one shall be, as before, under Government protection, and his employer answerable as before to Government for the way he is housed, treated when sick, &c., &c., it is not easy to see why such enactments are more necessary in his case than in that of any other hired servant or laborer throughout all India.

All evidence collected, all enquiries made, tend to show that coolies are well treated on Tea estates. It is to the interest of the proprietors and managers to do so, and self-interest is a far more powerful inducement than any the Government can devise. The meddling interference caused by the imported coolie laws in the visits of the "Protector of Coolies"* to a garden conduce to destroy the kind feelings which should (and in spite of these hindrances often do) exist between the proprietor or manager and his men. I do not hesitate in my belief that imported coolies on Tea plantations would be better off in many ways were all Government interference abolished.

I do not decry Government action to the extent of seeing the coolies understand their terms of engagements and are cared for on the journey to the Tea districts; but once landed on the garden all Government interference should cease.

The idea of the State laying down how many square yards of jungle each coolie shall clear in a day, how many

---

* What a designation! Who invented it, I wonder? A clever man, doubtless, for Government interference was probably his hobby, and he quickly perceived the very title would, more or less, render the office necessary!

square feet he shall dig, &c., &c., &c.! Can *any* certain rates be laid down for such work? Is all jungle the same, all soil the same, and even if such rates *could* be laid down how can the rules be followed? Bah! they are *not*, never will be, and the whole thing is too childish for serious discussion.

It is not difficult to sit at a desk and frame laws and rules that look feasible on paper. It is quite another thing to carry them out. Over-legislation is a crying evil in India, but there is still a worse, namely legislation and official action, on subjects of which the said officials are utterly ignorant.

I have said enough to show imported labor cannot vie with local, nor would it do so were all the evils of Government interference removed. I therefore believe Tea property in India will eventually pay best where local labor exists. This will naturally be the case when other conditions are equal, but so great are the advantages of local labor, I believe it will also be the case in spite of *moderate* drawbacks.

CHAPTER III.

TEA DISTRICTS AND THEIR COMPARATIVE ADVANTAGES, CLIMATE, SOIL, &c., IN EACH.

THE Tea districts in India, that is where Tea is grown in India to-day, are:—

1. Assam.
2. Cachar and Sylhet.*
3. Chittagong.
4. Terai below Darjeeling.
5. The Dehra Dhoon.
6. Kangra (Himalayas).
7. Darjeeling (Himalayas).
8. Kumaon (Himalayas).
9. Hazareebaugh.
10. Neilgherries (Madras Hills).

In fixing on any district to plant Tea in four things have to be considered, *viz.* soil, climate, labor, and means of transport. When the district being selected a site has to be chosen, all but the second of these have to be considered again, and the lay of land, nature of jungle, water, and sanitation are also of great importance in choosing a site.

I will first then discuss generally the Tea districts given above as regards the advantages of each for Tea cultivation. I have seen and studied Tea gardens in all the districts named, except Nos. 5 and 10. What I know of these two is from what I have read, and what is generally known of their climates.

---

* These are virtually one, and I shall allude to both as Cachar.

TEA DISTRICTS. 13

Before, however, comparing each district, we should know what are the necessities of the Tea-plant as regards climate and soil. Tea, especially the China variety, will grow in very varying climates and soils, but it will not flourish in all of them, and if it does not flourish, and flourish well, it will certainly not pay.

The climate required for Tea is a hot damp one. As a rule a good Tea climate is not a healthy one. The rainfall should not be less than 80 to 100 inches per annum, and the more of this that falls in the early part of the year the better. Any climate which though possessing an abundant rainfall suffers from drought in the early part of the year, is not *cæteris paribus* so good, as one where the rain is more equally diffused. All the Tea districts would yield better, with more rain in February, March, and April, and therefore some, where fogs prevail in the mornings at the early part of the year, are so far benefitted.

As any drought is prejudicial to Tea, it stands to reason hot winds must be very bad. These winds argue great aridity, and the Tea-plant luxuriates in continual moisture.

The less cold weather experienced where Tea is, the better for the plant. It can stand, and will grow, in great cold (freezing point, and lower in winter is found in some places where Tea is), but I do not think it will ever be grown to a profit on such sites. That Tea requires a temperate climate, was long believed and acted upon by many to their loss. The climate *cannot* be too hot for Tea if the heat is accompanied with moisture.

Tea grown in temperate climes, such as moderate elevations in the Himalayas, is quite different to the Tea of hot, moist climates, such as Eastern Bengal. Some people like it better, and I believe the flavor is more delicate; but it is very much weaker, and the value of Indian Tea (in the present state of the home market where it is principally used for giving "body" to the washy stuff from China) consists in its strength. Another all-important point in fixing on a climate for Tea is the fact, that apart from the strength the yield is double in hot, moist climes, what it is in comparatively dry and temperate ones. A really pleasant climate to live in *cannot* be a good one for Tea. I may now discuss the comparative merits of the different Tea districts.

ASSAM.

This is the principal home of the indigenous plant, and were it not for scarcity of labor no other district could vie with it. The climate in the northern portions is perfect, superior to the southern, as more rain falls in the spring. The climate of the whole of Assam, however, is very good for Tea. The Tea-plant yields most abundantly when hot sunshine and showers intervene. For climate then I accord the first place to Northern Assam. Southern Assam is as observed a little inferior.

The soil of this province is decidedly rich. In many places there is a considerable coating of decayed vegetation on the surface, and inasmuch as in all places where Tea has been or is likely to be planted it is strictly virgin soil, considerable nourishment exists. The prevailing soil also is light and friable, and thus, with the exception of the rich oak soil in parts of the Himalayas, Assam in this respect is second to none.

As regards labor we must certainly put it the last on the list. The Assamese, and they are scanty, won't work, so the planters, with few exceptions, are dependant on imported coolies, and inasmuch as the distance to bring them is enormous, the outlay on this head is large, and a sad drawback to successful Tea cultivation.

The Berhampootra—that vast river which runs from one end of Assam to the other—gives an easy mode of export for the

Tea, but still owing to the distance from the sea board it cannot rank in this respect as high as some others.

CACHAR.

The indigenous Tea is also found in a part of this province. The climate is a little inferior to Assam, because the rains are too heavy, but I think it takes the second place.

The soil is not equal to Assamese soil, it is more sandy, and lacks the power. Again, there is much more flat land fit for Tea cultivation in Assam, and there can be no doubt as to the advantage of level surfaces.

As regards transport Cachar has the advantage, for it has equally a water-way, and is not so distant from Calcutta.

The labor aspect is much the same in the two provinces, both being almost entirely dependant on imported coolies; but Cachar is nearer the labor fields than Assam.

However, after discussing separately the advantages of each province, I propose to draw up a tabular statement, which will show at a glance the comparative merits of each on each point discussed.

CHITTAGONG.

This is a comparatively new locality for Tea. The climate is better than Cachar in the one respect that there is less cold weather, but inferior in the more important fact that much less rain falls in the spring. In this latter respect it is also inferior to Assam, particularly to Northern Assam. I therefore as to climate give it the third place. There is one part of Chittagong, the Hill Tracts, (Tea has scarcely been much tried there yet) which, in the fact of spring rains, is superior to other parts of the province as also in soil, for it is much richer there. On the whole, however, Chittagong must yield the palm to both Assam and Cachar on the score of climate, and also I think of soil. For though good rich tracts are occasionally met with, they are not so plentiful as in the two last-named districts. Always, however, excepting the Hill Tracts of Chittagong, there the soil is, I think, quite equal to either Assam or Cachar.

As regards labor (a very essential point to successful Tea cultivation) Chittagong is most fortunate. With few exceptions (and those only partial) all the plantations are carried on with local labor, which excepting for about two months, the rice-time, is abundant.

For transport (being on the coast with a convenient harbour, a continually increasing trade, ships also running direct to and from England) it is by far the most advantageously situated of all Tea localities.

Chittagong possesses another advantage over all other Tea districts in its large supply of manure. The country is thickly populated, and necessarily large herds of cattle exist. The natives do not use manure for rice (almost the sole cultivation), and, consequently, planters can have it almost for the asking. The enormous advantages of manure in Tea cultivation are not yet generally appreciated. It will certainly double the ordinary yield of a Tea garden. A chapter is devoted to this subject.

Terai below Darjeeling.

I have seen this, and the Tea in it, since I wrote the first edition of this Essay.

The soil is *very* good for Tea. The climate is also a good one, but there is not as much rain in the early part of the year as planters could wish. Much difficulty exists about labour owing to the very unhealthy climate. As the jungle is cleared however, this last objection will be in a measure got over. As it stands now it is perhaps the most unhealthy Tea locality in India.

Communication will be very easy when the Northern Bengal Railway is finished.

Except in the point of salubrity (which is however an important one), I think this locality a favorable one for Tea.

THE DEHRA DHOON.

I have heard the first Tea in India was planted here. The lucky men, two officers, who commenced the plantation, sold it, I believe in its infancy, to a Company for five lakhs of rupees. What visions did Tea hold forth in those days!

In climate the Dehra Dhoon is far, far from good. The hot dry weather of the North-West is not at all suited to the Tea plant. Hot winds shrivel it up, and though it recovers when the rains come down, it cannot thrive in such a climate. One fact will, I think, prove this. In favorable climates, with good soil, and moderate cultivation, 18 flushes or crops may be taken from a plantation in a season. With like advantages, and *heavy* manuring, 22 or even more may be had. In the "Selections from the Records of the Government of India" on Tea published in 1857 (a book to which many owe their ruin), the following appears, showing how small are the number of flushes in the North-West:—

Method of gathering Tea Leaves.—"The season for gathering leaves generally commences about the beginning of April, and continues until October; the number of gatherings varies, depending on the moistness and dryness of the season. If the season be good, that is to say, if rain falls in the cold weather and spring, and the general rains be favorable, as many as five gatherings may be obtained. These, however, may be reduced to three general periods for gathering,

<small>Three general gatherings.</small>

viz. from April to June, from July to 15th August, and from September to 15th October. If the season be a dry one, no leaves ought to be taken off the bushes after the 1st October, as by doing so they are apt to be injured. If, however, there are good rains in September, leaves can be pulled until the 15th October, but no later, as by this time they have got hard and leathery and not fitted for making good Tea, and as it is necessary to give the plants good rest in order to recruit. Some plants continue to throw out new leaves until the end of November; but those formed during this month are generally small and tough."

When this was written, the experience detailed related to Dehra Dhoon, the Kumaon, and Kangra Gardens, and we see that five flushes or gatherings are thought good. It however makes matters in this respect (far from a general fault in the said " Records") worse than they are. Ten and twelve flushes, with *high* cultivation, can be got in the North-West. But what is this as against 20 and 25 ?

Labor is plentiful and cheap. The great distance from the coast makes transport very expensive.

KANGRA.

This is a charming valley, with a charming climate more favorable to Tea than the Dehra Dhoon, still it is far from a Tea climate. It is too dry and too cold. The soil is good for Tea, better than that of the Dhoon, but inferior to some rich soils in the Himalayan oak forests. Local labor is obtainable at cheap rates. Distance makes transport for export very difficult; but a good local market now exists in the Punjab, and a good deal of Tea is bought at the fairs, and taken away by the wild tribes over the border. With the limited cultivation there, I should hope planters will find a market for all their produce. Manure must be obtainable (manure had not been thought of for Tea when I visited Kangra), and if liberally applied, it will increase the yield greatly.

Kangra is strictly a Himalayan district, but the elevation is moderate, if I remember right, about 3,000 feet, and the land is so slightly sloping it may almost be called level. A great advantage this over the steep lands, on which most of the Himalayan gardens, many in Cachar, and some in Assam and Chittagong, are planted.

Kangra is *not* the place for a man who wants to make money by Tea; but for one who would be content to settle there, and content to make a livelihood by it, a more desirable spot with a more charming climate could not be found. Land, however, is not easily procured.

DARJEELING.

This, too, I have seen since I published the first edition of this Essay. The elevation of the station, 6,900 feet, is far too great; but plantations lower down do tolerably well (that is well for hill gardens). The climate, like all hill climates, is too cold. As regards transport the Darjeeling plantations will be well situated when the railroad now constructing is finished. Like elevations in Darjeeling and Kumaon are in favor of the former, *first*, because the latitude is less; *secondly*, because Darjeeling has much more rain in the spring. I believe, therefore, that the hill plantations of Darjeeling have a better chance of paying than the gardens in Kumaon, but, as stated before, no elevated gardens, that is none in the Himalayas, have any chance in the race against plantations in the plains, always providing the latter are in a good Tea climate.

In two respects, however, Darjeeling is behind Kumaon. The soil is not so good, and the land is much steeper. It is more than absurd, some of the steeps on which Tea is planted in the former, and such precipices can, I am sure, never pay even the cost of their yearly cultivation. Gardens, barely removed above the Terai (and there are such in Darjeeling), can scarcely be called "elevated," and for them the remarks applied to the Terai are more fitting. As a broad rule it should be recognized that the lower Tea is planted in the Himalayas the better chance it has.

All the plants in the Darjeeling gardens, with but few exceptions, are China.

The China plant makes by far the best Green Tea, and I believe the Darjeeling gardens would pay much better than they do if they altered their manufacture from black to green. (See further on under the head of Hazareebaugh what has been done in this way). All Himalayan gardens should, in my opinion, make Green Tea (Kumaon has awoke to the fact) for all have China plants, and can therefore make far better Green Tea than can be produced from the Hybrid which is so general in plain gardens.

They would then have a manufacture of their own in which they could excel.

'Tis true the Green Tea market is yet uncertain, and large quantities *may* bring down prices (which at present rule very high, much higher than for Black Tea); but that limit is far from reached yet, and I think the demand for Green Tea is increasing.

In two words, as Kumaon and Hazareebaugh have done so well with Green Tea, it is very strange Darjeeling, with all the necessities for it, does not follow their example.

KUMAON.

It was in this district (a charming climate to live in, with magnificent scenery to gaze at) I first planted Tea in India, and I much wish for my own sake, and that of others, I had not done so. I knew nothing of Tea at the time, and I thought a district, selected by Government for inaugurating the cultivation, must necessarily be a good one. No hill climate *can* be a good one for Tea; but the inner parts of Kumaon, very cold, owing to its elevation, high latitude, and distance from the plains, is a peculiarly bad one. Yet there it was Government made nurseries, distributed seed gratis, recommended the site for Tea (see the "Records" alluded to), and led many on to their ruin by doing so. The intention of the Government was good, but the officers in charge of the enterprize were much to blame, perhaps not for making the mistake at first (no one *at the first* knew what climate was suitable), but for perpetuating the mistake, when later, very little enquiry would have revealed the truth. I believe it was guessed at by Government officials long ago, but it was easier to sing the old tune, and a very expensive song it has proved to many.*

I need scarcely, after this, add I do not approve of Kumaon for Tea. An exhilarating and bracing climate for man is not suited to the Tea-plant. The district has one solitary advantage—rich soil. I have never seen richer, more productive land than exists in some of the Kumaon oak forests, but even this cannot in the case of Tea counterbalance the climate. Any crop which does not require much heat and moisture will grow to perfection in that soil. Such potatoes as it produces! Were the difficulties of transport not so great, a small fortune might be made by growing them.

Could any part of Kumaon answer for Tea it would be the lower elevations in the outer ranges of the hills, but these are precisely the sites that have *not* been chosen. Led, as in my own case, partly by the Government example, partly by the wish to be *out* of sight of the "horrid plains," and *in* sight of that glorious panorama, the snowy range, planters have chosen the interior of Kumaon. Some wisely (I was not one of them) selected low sites, valleys sheltered from the cold winds, but even their choice has not availed much. The frost in winter lingers longest in the valleys, and though doubtless the yield there is larger, owing to the increased heat in summer, the young plants suffer much in the winter.

---

* Is it possible that the continued deception (it was nothing less) was owing to the fact Government had gardens to sell there? They were advertised for sale a long time at absurd prices.

The outer ranges, owing to the heat radiating from the plains, are comparatively free from frost, but there again the soil is not so rich. Still they would unquestionably be preferable to the interior.

Labor is plentiful in Kumaon and very cheap, Rs. 4 per mensem. Transport is very expensive. It costs, not a little, to send Tea from the interior over divers ranges of hills to the plains. It has then some days' journey by cart, ere it meets the rail, to which 1,000 miles of carriage on the railroad has to be added.

Since the above was written Kumaon has secured a good local market, and I believe sells most of its Tea unpacked to merchants who come from over the border to buy it.

It has also improved its position greatly by making Green Teas, for which, as observed before, the China plant it has is so well fitted. With those two advantages, though the climate is inferior, I suspect that Tea there now pays better than in Darjeeling.

Gurhwall is next to Kumaon and so similar, I have not thought it necessary to discuss it separately. The climate is the same, the soil as a rule not so good. There is one exception though, a plantation near "Lohba," the Teas of which (owing I conceive to its peculiar soil) command high prices in the London market. The gardens, both in Kumaon and Gurhwall, have been generally much better cared for than those in Eastern Bengal. As a rule they are private properties managed by the owners.

HAZAREEBAUGH.

This district I have resided in since I wrote the first edition of this Essay. The climate is too dry, and hot winds are felt there. A great compensation though is labor, it is more abundant and cheaper in this district than in any other. The carriage is all by land, and it is some distance to the rail. But the Tea gardens at Hazareebaugh can never vie with those in Eastern Bengal, inasmuch as the climate is very inferior.

The soil is very poor.

In short Hazareebaugh is in no way a good place for Tea; and I have reason to believe the gardens there never paid at all until three years ago, when they altered the manufacture from black to green. All is China plant there, so they can, and do, make very good Green Tea. I believe they now average Rs. 1-3 to Rs. 1-4 per ℔. all round, whereas their average of Black Tea was perhaps 12 or 13 annas. The produce per acre, owing to climate, soil, and China plant, is small; and had they not turned to Green Tea, I do not see how the gardens there could have lasted.

NEILGHERRIES.

This I have never seen. The climate is perhaps superior to the Himalayan, for the frost is very slight. Were there, however, more heat there in summer it would be better.

Some of the Teas have sold very well in the London market, and I believe that, in this district too, Green Tea is made. As the plants there, I have been told, are China there can be no doubt that Green Tea will pay best.

Still what little I say here about the Neilgherries can in no way be depended on, for not only have I never seen them, I have been told very little about them.

I have heard the soil is good, but have no certain information on this head. Not much difficulty can exist in the way of transport.

Having now discussed each district, all of which, except the Dehra Dhoon and Neilgherries, I have seen, I give for further elucidation Meteorological Tables of the principal ones. For those not mentioned in the tables I have failed to acquire the necessary information.

My thanks are due to Dr. Coates at Hazareebaugh for his kindness in supplying me with much of the data from which the following tables are framed:—

Table of Elevation and Temperature of Tea Localities.

N.B.—The exact temperature of other Tea Districts not being known, I have confined myself to these; but general remarks on the elevation and temperature of other Tea localities will be found elsewhere.

Districts.	Place.	Elevation in feet.	Details.	January.	February.	March.	April.	May.	June.	July.	August.	September.	October.	November.	December.	D.J.F.	M.A.M.	J.J.A.	S.O.N.	Year.
Assam.	Goalparah ...	366	Monthly Temp.	61·7	63·0	72·6	77·6	78·0	80·3	83·1	81·6	80·5	77·5	69·0	64·6	63·1	75·4	81·3	75·8	73·8
			Do Max.	77·2	87·9	94·0	97·0	91·0	91·0	92·0	91·5	89·0	89·0	84·3	78·3					
			Do Min.	49·0	48·0	57·2	62·8	67·0	70·0	73·7	73·0	70·1	62·3	50·8	50·0					
	Gowhatty ...	131	Monthly Temp.	63·8	67·8	74·6	77·4	80·4	81·8	83·0	82·9	82·3	79·2	71·1	65·5	65·6	77·4	82·8	77·5	75·8
	Seebsaugor ...	370	Monthly Temp.	60·0	64·1	69·3	73·8	78·5	81·4	83·6	83·6	83·1	78·3	69·4	62·4	62·2	73·7	83·2	76·9	74·0
	Debrooghur ...	396	Monthly Temp.	62·2	63·4	71·3	72·7	77·1	80·7	83·7	81·7	81·0	75·0	67·4	61·0	62·2	73·7	82·1	74·7	73·2
Cachar.	Cachar ...	76	Monthly Temp.	62·9	66·6	73·4	76·8	80·9	82·2	83·3	81·7	81·2	70·6	70·6	65·2	64·9	77·0	82·4	77·1	75·3
Chittagong.	Chittagong ...	191	Monthly Temp.	69·5	72·3	80·5	83·5	84·5	84·0	82·2	82·3	83·0	81·6	73·7	66·9	69·9	82·8	82·8	79·4	78·7

Table of Elevation and Temperature of Tea Localities.—(Continued.)

N.B.—The exact temperature of other Tea Districts not being known, I have confined myself to these; but general remarks on the elevation and temperature of other Tea localities will be found elsewhere.

Districts.	Place.	Elevation in feet.	Details.	January.	February.	March.	April.	May.	June.	July.	August.	September.	October.	November.	December.	D. J. F.	M. A. M.	J. J. A.	S. O. N.	Year.
Darjeeling.	Darjeeling	6982	Monthly Temp.	43·2	43·8	52·0	58·7	63·1	63·7	64·1	64·4	63·0	57·3	49·4	44·7	43·5	57·8	64·3	55·4	55·9
			Do Max.	63·0	60·0	72·0	78·0	79·0	79·0	76·0	75·0	80·0	78·0	69·0	60·0					
			Do Min.	32·0	28·0	39·0	46·0	48·0	57·0	56·0	59·0	57·0	44·0	38·0	33·0					
Chota Nagpore.	Hazareebaugh	2010	Monthly Temp.	65·7	67·1	73·7	85·6	88·6	83·8	77·8	79·3	77·8	72·6	64·8	61·4	63·7	83·6	80·3	71·8	74·5
			Do Max.	82·0	91·0	94·0	107·0	100·0	103·0	89·0	88·0	87·0	84·0	78·0	78·0					
			Do Min.	44·0	46·4	55·0	67·0	73·0	71·0	71·0	73·0	70·0	59·0	52·0	44·0					
Neil-gherries.	Ootacamund	7480	Monthly Temp.	51·5	53·8	57·3	60·1	60·8	57·9	55·8	56·1	56·4	55·9	53·9	51·8	52·1	59·4	56·6	55·4	55·9

N.B.—The letters in the columns, between December and the year, refer to months; thus, D. J. F. in December, January, February. The figures show the average temperature during those months.

Table of Latitude, Longitude, and Rainfall of Tea Localities.

N.B.—The exact rainfalls of other Tea Districts not being known, I have confined myself to these; but general remarks on the rainfall in other Tea localities will be found elsewhere.

Districts.	Place.	Latitude.	Longitude.	Detail.	January.	February.	March.	April.	May.	June.	July.	August.	September.	October.	November.	December.	Totals.
Assam.	Goalparah	26°11′	90°36′	Average rain, several years	0·43	0·78	1·84	4·95	11·72	23·72	21·33	12·69	10·93	5·91	0·39	0·20	94·44
				Days rain fell in 1869	2	2	4	8	19	24	22	18	15	5	Nil	Nil	119
Assam.	Gowhatty	26°5′	91°43′	Average rain, several years	0·70	1·43	1·46	7·27	10·92	13·29	13·06	11·98	6·82	3·20	0·47	0·12	70·76
				Days rain fell in 1869	2	2	4	8	15	16	9	10	14	2	Nil	1	84
Assam.	Seebsaugor	27°2′	94°39′	Average rain, several years	1·18	2·43	3·77	10·15	11·04	15·66	14·97	13·88	11·13	4·46	1·29	0·09	90·45
				Days rain fell in 1869	11	9	10	13	22	13	19	23	17	8	Nil	2	147
Cachar.	Cachar	24°48′	92°43′	Average rain, several years	0·50	3·63	6·09	12·69	16·12	19·65	21·58	16·84	13·90	7·77	7·03	0·79	123·3
				Days rain fell in 1869	2	9	10	16	18	20	19	25	19	8	Nil	Nil	145
Chitta-gong.	Chittagong	22°20′	91°45′	Average rain, several years	0·37	1·82	1·31	5·46	9·43	22·92	22·54	23·04	13·01	5·98	2·30	0·65	108·47
				Days rain fell in 1869	1	7	3	4	14	15	21	25	17	5	Nil	1	113

Table of Latitude, Longitude, and Rainfall of Tea Localities.—(Continued.)

N.B.—*The exact rainfalls of other Tea Districts not being known, I have confined myself to these; but general remarks on the rainfall in other Tea localities will be found elsewhere.*

Districts.	Place.	Latitude.	Longitude.	Detail.	January.	February.	March.	April.	May.	June.	July.	August.	September.	October.	November.	December.	Total.
Chittagong.	Hill Tracts ...	?	?	Rain in 1869 ...	Nil	1·00	1·50	12·55	9·0	12·50	18·20	14·30	12·70	5·70	Nil	0·30	88·95
				Days rain fell in 1869	Nil	4	4	7	13	16	22	19	19	4	Nil	1	109
Darjeeling.	Darjeeling ...	27° 3′	89° 18′	Average rain, several years	0·75	1·50	1·85	3·62	7·01	27·50	29·40	29·09	18·06	6·56	0·20	0·14	129·59
				Days rain fell in 1869	2	8	5	9	17	23	26	22	24	7	1	2	148
	Buxar { Bhootan	?	?	Rain in 1869 ...	0·80	2·00	1·50	6·90	25·30	27·30	46·50	85·50	46·50	9·90	P	2·40	252·00
	Dooars			Days rain fell in 1869	3	3	6	7	15	19	25	28	22	5	P	P	P
Chota Nagpore.	Hazareebaugh ...	24° 0′	85° 20′	Average rain, several years	0·42	0·52	0·75	0·42	1·37	10·99	14·63	11·44	6·26	3·51	0·19	0·03	50·52
				Days rain fell in 1869	4	Nil	7	Nil	5	11	24	16	21	9	Nil	1	98

TEA DISTRICTS.

I will now endeavour to draw up a tabular statement of the respective advantages of the various Tea districts as regards climate, labor, lay of land, soil, facilities of procuring manure and transport.

In importance I regard them in the order given. I place labor before soil, because the fact is in all the provinces suitable and good soil for Tea can be found *somewhere;* and therefore while soil is all important in selecting a site, it is secondary to labor in deciding on a district. Lay of land comes after labor. When my information on any point is not sure I place a note of interrogation. Where advantages are equal, or nearly so, I give the same number, and the greater the advantage of a district on the point treated in the column the smaller the number. Thus under the head of climate Assam is marked 1, meaning it is the best.

As the following table gives no information as to which of all the districts possesses the greatest advantages, *all things considered,* but only gives my opinion of each under each head, and the subject closed in this way would be unsatisfactory, I may state that, in my opinion, the choice should lie between the three first on the list.

Comparative advantages of the Tea Districts in India as regards climate, labor, lay of land, soil, manure, and transport.

Tea Districts.	Climate.	Labor.	Lay of Land.	Soil.	Manure.	Transport.	
Assam	1	4	1	1	4	3	Water carriage.
Cachar	2	4	2	2	4	2	
Chittagong	3	2	2	2	1	1	
Chittagong Hill Tracts	3	3	3	1	2	1	
Terai below Darjeeling	2	4	1	1	3	5	Land carriage.
Darjeeling	4	3	5	3	3	6	
Hazareebaugh	6	1	1	4	2	4	
Kangra	4	3	1	3	3	9	
Debra Dhoon	5	3	1	3	3	7	
Kumaon	5	3	4	2	3	8	
Nilgherries	?	?	?	?	?	4	

CHAPTER IV.

SOIL.

To pronounce as precisely on soil as on climate is not easy. The Tea plant will grow on almost any soil, and will flourish on many. Still there are broad general rules to be laid down in the selection of soils for Tea, which no one can ignore with impunity.

When first I turned my attention to Tea, I collected soils from many gardens, noting in each case how the plants flourished. I then set down to examine them, never doubting to arrive at some broad practical conclusions. I was sadly disappointed. I found the most opposing soils nourished, apparently, equally good plants. I knew not then much about Tea, and judged of the Tea bushes mostly by their size (a very fallacious test); still after-experience has convinced me, I was more or less right in the conclusion I then came to, that several soils are good for Tea.

Nothing then but broad general rules can be laid down on this point, for I defy any one to select any one soil, as the best for Tea, to the exclusion of others. A light sandy loam is perhaps as good a soil as any out of the Himalayas. It ought to be deep, and the more decayed vegetable matter there is lying on its surface the better. If deep enough for the descent of the tap-root, say three feet, it matters not much what the sub-soil is, otherwise a yellowish red sub-soil is an advantage. This sub-soil is generally a mixture of clay and sand. Much of Assam, Cachar, and Chittagong is as the above, but as a rule it is richest in Assam, poorest in Chittagong.

Where the loam is of a greasy nature (very different to clay) with a mixture of sand in it, it is superior to the above, for it has more body. All good Tea soils must have a fair proportion of sand, and if not otherwise apparent, it may be detected by mixing a little of the soil with spittle, and rubbing it on the hand. If the hand is then held up towards the sun, the particles of sand will be seen to glisten.

The soil so common in Kumaon that is light rich loam with any amount of decayed vegetable matter on it, and with a ferruginous reddish yellowish sub-soil, is, I consider, the finest soil in the world for Tea. What a pity it nowhere exists in a really good Tea climate. The rich decayed vegetable matter is the produce for centuries of the oak leaves in the Himalayan forests, and as all the world knows oak only grows in temperate climes.

It was long believed that Tea would thrive best on poor soil. The idea was due to the description of Tea soils in China to be found in the first books that treated of Tea. But the fact that Tea, as a rule, is only grown in China on soil which is useless for anything else quite alters the case. If a soil is light and friable enough it cannot be too rich for Tea.

Ball's book "on the cultivation and manufacture of Tea in China" has much on Tea soils, but the opinions the author collected are sadly at variance, and on the whole teach nothing.

In conclusion I will attempt to point out the qualities in soils in which the Tea-plant delights, as also the qualities it abhors.

It loves soils friable, that is easily divided into all their atoms. This argues a fair proportion of sand, but this should not be in excess, or the soil will be poor. The soil should be porous, imbibing, and parting with water freely. The more decayed vegetable matter on its surface the better.

SOIL.

To be avoided are stiff soils of every kind, as also those which when they dry, after rain, cake together and split. Avoid also black colored, or even dark colored earths. All soils good for the Tea-plant are light colored. If, however, the dark color arises from decayed vegetation that is not the color of the soil, and, as observed, vegetable matter is a great advantage. Judge of color when soil is dry—for even light-colored soil looks dark when wet. Soil which will make bricks will not grow tea, and though I have sometimes seen young plants thrive on stiff soil, I do not believe in any stiff soil as a permanence.

Stones, if not in excess, are advantageous in all soils inclined to be stiff, for they help to keep them open. But then they must not be large, as if so they act as badly as a rocky substratum preventing the descent of the tap-root.

The reason, I take it, why Tea thrives best in light soils is that the spongioles or ends of the feeding roots are very tender, and do not easily penetrate any other.

There is more nourishment in stiffer soils, but for this reason the Tea-plant cannot take advantage of it.

If a chosen soil is too stiff it may be much improved for Tea by mixing sand with it. However, even where sand is procurable near, the expense of this is great. When done, the sand should be mixed with the soil taken out of the holes in which the plants are to be placed (see transplanting), and it may be done again later by placing sand round the plants and digging it in. All this though is extra labor and very expensive, so none but a good Tea soil should ever be selected, and it is very easily found, for it exists in parts of all the districts discussed.

CHAPTER V.

NATURE OF JUNGLE.

I HAVE not much to say under this head. I have heard many opinions as to the kind of trees and jungle that should exist in contemplated clearances, but I attach little or no weight to them, at all events in Bengal.

In the Himalayas it is somewhat different. There oak trees should be sought for, their existence invariably makes rich soil.* Fir on the contrary indicates poor soil. At elevations, however, the desideratum of a warm aspect interferes, for the best oak forests are on the colder side. I speak of course of elevations practicable, say three or four thousand feet, above this it is a waste of money to try and cultivate Tea.

In Bengal I do not think the nature of the jungle on land contemplated signifies much. As a rule, the thicker the jungle the richer the soil; but if seeking for a site, large trees should not be a *sine qua non*. Much of the coarse grass land is very good, and large trees add enormously to the expense of clearings. It is not cutting them down which is so expensive, it is cutting them up and getting rid of them by burning, or otherwise, after the former is done.

I have discussed soil fully already, and need only add here that if the knowledge to do so exists, it is better to judge of soil from the soil itself than from the vegetation on it, though doubtless a fact that luxuriant vegetation indicates rich soil.

---

* The oak tree leaves cause a rich deposit of vegetable matter.

CHAPTER VI.

WATER AND SANITATION.

THESE may be discussed together and shortly. Of course adjacent water carriage is a great advantage for a garden, and it should be obtained, if possible, in selecting a site. The expense of land carriage, Tea being such a bulky article, is great, and Tea cultivation requires all advantages to make it pay well.

But it is water for a garden that particularly concerns us now. It is not easy to find land that can be irrigated (this is discussed elsewhere), but no labor or expense in getting such land would be thrown away. Irrigation, combined with high cultivation in other respects, will give a yield per acre undreamt of.

In no case should a plantation be made except where a running stream is handy. Water is a necessity for seedlings, and a plentiful adjacent supply of it is a great desideratum for the comfort and health of every soul on the garden. We all know how dependent the natives are on water, and it is evident facilities in this respect will conduce much (whether the labor be local or imported) both to get and keep coolies.

It has been observed that, as a rule, a good Tea climate is not a healthy one. There is no getting over the fact, and we can only make the best of it. The house, the factories, and all the buildings should be placed as high as possible, and not very close to each other, both for the sake of health and in the event of fire. The locality should be well drained, and cleanliness be attained in every possible way. Give the coolies good houses, with raised mechans to sleep on, and sprinkle occasionally carbolic acid powder in your own house, and those of others.

Sanitation is however a large subject. It can be studied elsewhere. General ideas on it and on the properties of the commonest medicines are a great advantage to any intending Tea-planter.

CHAPTER VII.

LAY OF LAND.

The first idea prevailing about Tea was that it should be planted on slopes. It was thought, and truly, that the plant was impatient of stagnant water, and so it is, but it is not necessary to plant it on slopes in consequence. Pictures of Chinese, suspended by chains, (inasmuch as the locality could not be otherwise reached) picking Tea off bushes growing in the crevices of rocks somewhat helped this notion; and when stated, as it was, that the Tea produced in such places was the finest and commanded the highest price, intending planters in India went crazy in their search for impracticable steeps! Much of the failure in Tea has arisen from this fact, for a great part of many, the whole of some gardens, has been planted on land, so steep, that the Tea can never last or thrive on it. This is especially the case in Darjeeling.

Sloping land is objectionable in the following respects. It cannot be highly cultivated in any way, (I hold Tea will only pay with high cultivation) for high cultivation consists in frequent digging, to keep the soil open, and get rid of weeds and liberal manuring. If such soil is dug in the rainy season, it is washed down to the foot of the hill, and if manure is applied at any time of the year, it experiences the same fate when the rains come. As it cannot be dug, weeds necessarily thrive and diminish the yield by choking the plants.

The choice is therefore of two evils, "low cultivation and weeds" or "high cultivation which bares the roots of the plants in a twelve month." Of the two, the first *must* be chosen, for if the latter were pursued the plants getting gradually more and more denuded of soil, would simply topple over in two or three years. But choosing the lesser evil, the mischief is not confined to the bad effects of low cultivation. Dig the land as little as you will, the great force of the rains washes down a good deal of soil. The plants do not sink, as the soil lowers, and the consequence is that all Tea-plants on slopes have the lower side bare of earth, and the roots exposed. This is more and more the case the steeper the slope. These exposed roots shrivel up, as the sun acts on them the plant languishes, and yields very little leaf. Attempts are made to remedy the mischief by carrying earth up from below yearly, and placing it under the plant, but the expense of doing this is great, and the palliation is only temporary, for the same thing occurs again and again as each rainy season returns.

The mischief is greater on stiff than on sandy soils, for on the former the earth is detached in great pieces and carried down the hill. I know one garden in Chittagong, a large one, where the evil is so great, that the sooner the cultivation is abandoned the better for the owners.

A great many gardens in India, indeed the majority, are on slopes. A few in Assam, the greater number in Cachar, some in Chittagong, and almost all the Himalayan plantations. Such of these as are on *steep* slopes will, I believe, never pay, and instead of improving yearly (as good gardens, highly cultivated, should do even after they have arrived at full bearing) such, I fear, will deteriorate year by year.

Plantations on moderate slopes need not fail, because of the slopes. The evils slight slopes entail are not great, but the sooner the fact is accepted that sloping cannot vie against flat land for the cultivation of Tea the better.

Where only the lower part of slopes are planted, the plants do very well. The upper part being jungle the wash is not great, and the plants benefit much by the rich vegetable matter the rain brings down from above. I have often seen very fine plants on the lower part of slopes, where the upper has been left in jungle, and I should not hesitate to plant such portions *if* the slope was moderate.

Where teelah land, in Eastern Bengal, or sloping land in the Himalayas, Chittagong, or elsewhere, has to be adopted, aspect is all important. A good aspect in one climate is bad in another. In Assam, Cachar, Chittagong, and all warm places, choose the coolest, at high elevations (temperate climes) the warmest.

In the Himalayas, moreover, the warmer aspects are, as a rule, the most fertile; *vice versâ* in warm localities. Many a garden, which would have done very well, on the moderate slopes chosen, had *only* the proper aspects been planted, has been ruined by planting all sides of teelahs or hills indiscriminately. The southern and western slopes of plantations in warm sites are generally very bare of plants. Not strange they should be so, when the power of the reflected rays of the afternoon sun are considered. Again, in cold climates plants cannot thrive on northern aspects, for their great want in such climes is heat and sunshine. Let the above fault then be avoided in both cases, for though, doubtless, a garden is more handy, and looks better in one piece planted all over without any intervening jungle, even patches of jungle look better, and are decidedly cheaper, than bare cultivated hills.

Of flat land, after what I have written, I need not add much. It is of two kinds, table and valley land; the former is very rare in Tea districts, at least of any extent, which makes it worth while to plant it. There are two gardens in Chittagong on such flat table land, and they are both doing very well. Table land cannot be too flat, for the natural drainage is so great, no stagnant water can lie. It is inferior to valley land in the dry season, but superior in the rains.

Valley land is not good if it is *perfectly* flat. It will then be subject to inundation and stagnant water. There is nothing that kills the plant so surely and quickly as the latter. Even quite flat valleys can be made sweet by artificial drainage, but to do this a lower level, not too far distant, must exist, and the danger is not quite removed then. Valleys in which no water-course exists, and which slope towards the mouth alone, are to be avoided, for the plants near the mouth always get choked with sand. The best valleys are those with a gentle slope both ways, one towards the lowest line of the valley, be it a running water-course, or a dry nullah which carries off rain, the other towards the mouth of the valley. Such valleys drain themselves, or at least very little artificial drainage is necessary. A valley of this kind, with a running stream through it, is *most* valuable for Tea, and if the other advantages of soil and climate are present it is simply a perfect site. Such however are not frequent. If in such valleys, as is generally the case, the slope from the head to the mouth is enough, the running stream can be "bunded" (shut up) at a high level, and brought along one side at a sufficient elevation to irrigate the whole.

I have never seen but one garden in a valley that fulfils all these conditions exactly. It is in Chittagong, the soil is good, labor plentiful, and manure abundant. It ought to do great things, for the possibility of irrigating plants in the dry season (which as observed is very trying in Chittagong) will give several extra flushes in the year.

Of course in the wet season on such land the water must be allowed to resume its natural course.

Narrow valleys are not worth planting if the hills on the sides are steep, and they are consequently better left in jungle. No narrow tracts of land, with jungle on both sides, are worth the expense of cultivation, for the continual encroachment of the jungle gives much extra work. The plants, moreover, in very narrow valleys get half buried with soil, washed down from the adjacent slopes. Narrow valleys are therefore, in any case, better avoided.

To conclude, shortly, flat lands can be highly cultivated, steep slopes cannot. Tea pays best (perhaps not at all otherwise) with high cultivation, *ergo* flat lands are preferable.

CHAPTER VIII.

LAYING OUT A GARDEN.

BY this I mean, so dividing it when first made into parts, that later the said parts shall be easily recognized, and separately or differently treated, as they may require it.

The usual custom is to begin at one end of a plantation, and dig it right through to the other. In the same way with the pruning and plucking, and I believe the system is a very bad one. Different portions of gardens require different treatment, inasmuch as they differ in soil, and otherwise. One part of a plantation is much more prolific of weeds than another—how absurd that it should be cleaned no oftener! This is only one exemplification of difference of treatment, but in many ways it is necessary, most of all in plucking leaf.

All parts of a plantation, owing in some places to the different ages of the plants, in others to the variety in the soil and its productive powers, in others to slopes or to aspect do not yield leaf equally, that is, flush does not follow flush with equal rapidity. In some places (supposing each part to be picked when the flush is ready) seven days interval will exist between the flushes, in others nine, ten, or twelve, but no attention as a rule is paid to this. The pickers have finished the garden at the west end, the east end is again ready, and when done, the middle part will be taken in hand, be it ready, or be it not! It may be that the middle part flushes quicker than any other; in this case the flush will be more than mature when it is taken, in fact it will have begun to harden, or it may be the middle part does *not* flush as quickly as the others; in this case it will be picked before it is ready, that is, when the flush is too young, and the yield will consequently be smaller.

I believe the yield of a plantation may be largely increased by attending to this. Every Tea estate should be divided into gardens, of say, about six to ten acres each. If no natural division exists small roads to act as such should be made. More than this cannot be done when the plantation is first laid out, but when later the plants yield any difference between the productive powers of different parts of the same garden should be noted, and these divided off into sections. To do this latter with roads would take up too much space, and small masonry pillars, white-washed, are the best. Four of these, one at each corner of a section, are enough, and they need not be more than three feet high and a foot square. Thus each garden may where necessary be divided into two sections, which in a three hundred acre estate, partitioned off into 30 gardens, would give about 40 to 60 sections. No matter where a section may be, directly the flush on it is ready it should be picked. Where the soil on any one garden is much the same, and observation shows the plants all over it flush equally, it may be left all in one. I only lay down the principles, and I am very certain it works well, the proof of which is that where I have practised it some sections during the season give three, four, and five flushes more than others. Had the usual plan of picking from one end to the other been adopted, they would have been all *forced* to give the same number. In other words the said extra flushes would have been lost, and further loss occasioned by some flushes being taken before they were ready, others after a portion of the tender leaf had hardened.

The best plan is simply to number the gardens from one upwards, and the sections in each garden the same way. Thus supposing No. 5 garden is divided into three sections, they will be known respectively as 5-1, 5-2, and 5-3. This is the best way for the natives, and I find they soon learn to designate each section. I have a man whose special duty, (though he has other work also) it is to see each day which sections are ready to pick the following, and those, and those alone, are picked. Practice soon teaches the number of pickers required for any given number of sections, and that number only are put to the work. If a portion is not completed that day, it is the first taken in hand the next, and if any day on no sections is the flush ready, no leaf is picked the following.

Apart from leaf-picking, the garden and section plan detailed is useful in many ways. Each garden, if not each section which most requires it, is dug, pruned, or manured at the best time, and any spot on the plantation is easily designated. The plan facilitates the measurement of work and enables correct lists of the flushes gathered to be kept. It is thus seen which gardens yield best, and the worst can, by extra manuring, be brought to equal those.

In short the advantages are many, too numerous to detail.

Of course all this can be better done on a flat garden than on one planted on slopes, and though it may not be possible to work it out as much in detail on the latter, still a good deal in that way can be done, and I strongly recommend it.

In laying out a plantation keep it all as much together as possible, the more it is in one block the easier it is supervised the cheaper it is worked. Still do not, with a view to this, take in any bad land, for bad land will never pay.

Let your lines of Tea plants, as far as practicable, run with geometrical regularity. You will later find, both in measuring work and picking leaf, great advantages therefrom. In gardens where the lines are not regular portions are continually being passed over in leaf-picking, and thereby not only is the present flush from such parts lost, but the following is also retarded.

If your different gardens are so situated that the roads through them, that is, from one garden to the other, can be along *the side* of any garden without increasing the length of the road by all means adopt that route. There is no such good boundary for a garden as a road that is being continually traversed. It will save many rupees by preventing the encroachment of jungle into a garden, and more space is thus also given for plants. It is, however, of no use to do it if a road through the middle of the garden is shorter, as coolies *will* always take the shortest route.

The lines of plants on sloping ground should neither run up and down, nor directly across the slope. If they run up and down gutters or water-courses will form between the lines, and much additional earth will be washed away thereby. If they run right across the hill the same thing will occur *between the trees in each line,* and the lower side of each plant will have its roots laid very bare. It is on all slopes a choice of evils, but if the lines are laid diagonally across the hill, so that the slope *along the lines* shall be a moderate one, the evil is reduced as far as it can be by any arrangement of the plants. No, I forgot, there is one other thing. The closer the lines to each other, and the closer the plants in the lines to each other, in short the more thickly the ground on slopes is planted the less will be the wash, for stems and roots retain the soil in its place, and the more there are the greater the advantage.

Where slopes are steep (though remember steep slopes are to be avoided), terracing may be resorted to with advantage as the washing down of the soil is much checked by it.

On flat land of course it does not really signify in which directions the lines run, but such a garden looks best if, when the roads are straight, the lines run at right angles to them.

In laying out a garden choose a central spot with water handy for your factory, bungalow, and all your buildings;

let your Tea-houses be as close to your dwelling-house as possible, so that during the manufacturing time you can be in and out at all hours of the day and night. Much of your success will depend upon this. Let all your buildings be as near to each other as they can, but still far enough apart, that any one building may burn without endangering others. You need not construct any Tea buildings until the third year.

CHAPTER IX.

VARIETIES OF THE TEA-PLANT.

These are many, but they all arise from two species, the China plant, the common Tea bush in China, and the indigenous plant, first discovered some forty years ago in Assam.

These are quite different species of the same plant. Whether the difference was produced by climate, by soil, or in what way, no one knows, and here we have only to do with the facts that they *do* differ in every respect. A purely indigenous plant or tree (for in its wild state it may more properly be called the latter) grows with one stem or trunk and runs up to 15 and 18 feet high. It is always found in thick jungle and would thus appear to like shade. I believe it does when young; but I am quite sure if the jungle were cleared round an indigenous Tea tree found in the forest, it would thrive better from that day. The China bush (for it is never more) after the second year has numerous stems, and 6 or 7 feet would seem to be its limit in height. The lowest branches of a China plant are close to the ground, but in a pure cultivated indigenous, nine inches to a foot above the soil, up to which the single stem is clean.

The indigenous grows quicker after the second or third year than the China, if it has not been over-pruned or over-plucked when young. In other words, it flushes quicker, for flushing is growing.

The indigenous does not run so much to wood as the China. Indigenous seedlings require to be watered oftener than China, for the latter do not suffer as quickly from drought. The indigenous tree has a leaf of nine inches long and more: the leaf of the China bush never exceeds four inches. The indigenous leaf is a bright pale green, the China leaf a dull dark green color. The indigenous "flushes," that is, produces new tender leaf much more copiously than China, and this in two ways: *first*, the leaves are larger, and thus if only even in number exceed in bulk what the China has given; and *secondly*, it flushes oftener. The infusion of Tea made from the indigenous species is far more "rasping" and "pungent" than what the China plant can give, and the Tea commands a much higher price. It is difficult to prune the China plants too young, the indigenous on the contrary requires tender treatment in this respect. The young leaves from which alone Tea is made are of a much finer and softer texture in the indigenous than in the China, the former may be compared to satin, the latter to leather. The young leaves of the indigenous moreover do not harden so quickly as those of the China, thus if there is any unavoidable delay in picking a flush, the loss is less with the former. In the fact that unpruned or unpicked plants (for picking is a miniature pruning) give fewer and less succulent young leaves which harden quicker than pruned ones, the two varieties would seem to be alike. The China variety is much more prolific of seed than the indigenous, the former also gives it when younger, and as seed checks leaf, the China is inferior in this as in other respects. The China is by far the hardier plant. It is much easier to rear and it will grow in widely differing climates, which the indigenous will not.

A patch of indigenous with a mature flush on it is a pretty sight. The plants all appear as if crowned with gold (they are truly so if other advantages exist) and are a great contrast to the China variety if it can also be seen near.

I have now, I think, pointed out the leading characteristics of the two original varieties of the Tea-plant, and it stands to reason no one would grow the China who could get indigenous. But the truth is a pure specimen of either is rare. The plants between indigenous and China are called "hybrids." They

were in the first instance produced by the inoculation when
close together, of the pollen of one kind into the flower of the
other, and the result was a true hybrid, partaking equally of
the indigenous and China characteristics; but the process was
repeated again and again betweeen the said hybrid and an
indigenous or China, and again later between hybrids of
different degrees, so that now there are very many varieties of
the Tea-plant, 100 or even more, and no garden is wholly indigenous or wholly China. So close do the varieties run, no
one can draw the line and say where the China becomes a
hybrid, the hybrid an indigenous. Though as a rule the young
leaves are light green, or dark green, as the plant approaches,
the indigenous or China in its character, there are a certain
class of bushes (all hybrid) whose young leaves have strong
shades of crimson and purple. Some even are quite red, others
quite purple. These colors do not last as the leaf hardens,
and the matured leaves of these plants do not differ from
others. Plants with these colored leaves are prolific.

The nearer each plant approaches the indigenous the higher
its class and excellence, *ergo* one plantation is composed
of a much better class of plants than another. Had China
seed never been introduced into India a very different
state of things would have existed now. The cultivation
would not have been so large, but far more valuable. The
propagation and rearing of the indigenous as observed is difficult, the China is much hardier while young. So difficult is
it to rear successively the *pure* indigenous, perhaps the best
plan, were it all to come over again, would be to propagate
a high class hybrid and distribute it, never allowing any
China seed or plants to leave the nursery, which should
have been a Government one. But we must take things
as they are. The Government nurseries in the Himalayas
and the Dehra Dhoon (there have never been any elsewhere
and worse sites could not have been chosen) were planted

entirely with China seed, the seedlings distributed all over the country, and thus the mischief was done. The Indian Tea is vastly superior to China, and commands a much higher price at home, but it is still very inferior to what it would have been, had not China seed been so recklessly imported and distributed over the country.

The home of the indigenous Tea tree is in the deep luxurious jungles of Assam and Cachar.* There it grows into a good-sized tree. I have seen it 20 feet high. These are of no use, except for seed, until they are cut down. When this is done they throw out many new shoots, covered with young tender leaves, fit for Tea. They are of course far too big to transplant, but on some sites where they were numerous, that spot was chosen for the plantation, and some of these are the best gardens in Assam and Cachar.

The indigenous plant and high class hybrid requires a hot, moist climate, and will not therefore flourish in any parts of India outside Eastern Bengal. I have tried it in the Himalayas, there the cold kills it. In Dehra Dhoon and Kangra the climate is far too dry; besides the hot winds in the former, and the cold in the latter, are prejudicial. The Terai under Darjeeling would suit it. In Assam, Cachar, and Chittagong, the indigenous and the highest class hybrids will thrive, for the climate of all three suit it, but perhaps Northern Assam possesses the best climate of all, for that description of plant.

The Himalayan gardens consist entirely of China plants mixed occasionally with a low class of hybrid. They were all formed from the Government Nurseries where nothing but China was reared. Occasional importations of Assam and Cachar seed will account for the sprinkling of low class hybrids which may be found. The same may be said

---

* It is a singular fact that none exist in Northern Cachar, that is, on the northern side of the river.

of Dehra Dhoon and Kangra. In some gardens in the Terai below Darjeeling a high class of plant exists. In Assam, Cachar, and Chittagong the plantations vary much, but all have some indigenous and high class hybrids, while many gardens are composed of nothing else.

It is evident then that the value of a garden depends much on the class of its plants, and that a wise man will only propagate the best. Only the seed from good varieties should be selected, and gradually all inferior bushes should be rooted out and a good kind substituted. When this shall have been systematically done for a few years on a good garden, which has other advantages, the yield per acre will far exceed anything yet realised or even thought of.

Government action in the matter of Tea has been prejudicial in many ways, but in none more so than when they were doing their best to foster the cultivation by distributing China seed and seedlings gratis. No one can blame here (would the Government were equally free from blame in all Tea matters), but the mischief is none the less. It will take years to undo the harm then done.

The seed of indigenous, hybrid, and China are like in appearance and cannot be distinguished. Thus, when seed formerly was got from a distance, the purchaser was at the mercy of the vendor.

High cultivation improves the class of a Tea-plant. Thus, a purely China bush, if highly cultivated and well manured, will in two or three years assume a hybrid character. High cultivation will therefore improve the class of *all* the plants in a garden; but the cheapest and best plan with low class China plants is to root them out and replace them with others, as will be explained hereafter. Low class seedlings should also be rooted out of nurseries.

I cannot conclude this Chapter better than by giving an extract from the "Government Records" alluded to, and I add a few remarks at foot, as otherwise the reader might be puzzled with some opinions expressed so much at variance with the generally received opinions on Tea to-day.

Kinds of Tea-plants cultivated.—" When Government resolved on trying the experiment of cultivating Tea in India, they deputed Dr. Gordon to China to acquire information respecting the cultivation and manufacture of Teas, and to procure Tea seeds. Aided by Dr. Gutzlaff he procured a quantity of seeds from the mountains in the Amoy districts. These seeds were sent to the Calcutta Botanical Garden, where they were sown in boxes. On germinating they were sent up the country in boats, some to Assam and some to Gurhmuktesur, and from thence to Kumaon and Gurhwal. From these plants date the commencement of the Tea plantations in the Himalayas.* Tea was first made in Kumaon in 1841, and the samples sent to England, and were pronounced to be of good quality fitted for the home markets and similar to the Oolong Souchong varieties. Thus Messrs. Thompson, of Mincing Lane, report on a sample sent by us to Dr. Royle in 1842: "The samples of Tea received belongs to the Oolong Souchong kind, fine flavored and strong. This is equal to the superior black Tea generally sent as presents, and better for the most part than the China Tea imported for mercantile purposes." † By many it was supposed that there were different species of the Tea-plant, and that the species cultivated in the south districts of China was different from that met with in the north. To solve this mystery, and at the same time procure the best varieties of the Tea-plant, Mr. Fortune was deputed to China. By him large numbers of Tea-plants were sent from different districts of China celebrated for their Teas, and are now thriving

---

* And also the introduction of a bad class of plants.

† A single small sample of Tea very carefully made, and with an amount of labor which could never be bestowed on the mass, is little or no criterion. Tea is better made in Kumaon in 1874 than it was in 1842, but Kumaon Tea does not vie in price with Eastern Bengal produce. All the Himalayan Tea is weak, though of a delicate flavor; all Tea grown at high elevation *must* be so.

VARIETIES OF THE TEA-PLANT.

luxuriantly in all the plantations throughout the Kohistan of the North-West Provinces and Punjab. Both green and black Tea-plants were sent, the former from Whey Chow, Mooyeen, Chusan, Silver Island, and Tein Tang, near Ningpo, and the latter from Woo-e San, Tein San, and Tsin Gan, in the Woo-e district. But so similar are the green and black Tea-plants to each other, and the plants from the Amoy districts, that the most practised eye, when they are mixed together, cannot separate them,

Several varieties.

showing that they are nothing more than mere varieties of one and the same plant, the changes in the form of the leaf being brought about by cultivation. Moreover, throughout the plantation fifty varieties might easily be pointed out; but they run so into each other as to render it impossible to assign them any trivial character; and the produce of the seed of different varieties do not produce the same variety only but several varieties, proving that the changes are entirely owing to cultivation; nor do the plants, cultivated at 6,000 feet in the Himalayas, differ in the least in their varieties from those cultivated at 2,500 feet of altitude in the Dehra Dhoon.

"That the Assam plant is a marked species is true, it being distinguished by its large membraneous

Assam species.

and lanceolate leaf, small flower, and upright growth.

"It is a very inferior plant for making Tea, and its leaves are therefore not used.* Though the plants, received from the different districts of China do not differ from those first sent to the plantations, it is highly important to know that the Tea-plants from well-known green and black Tea districts of China now exist in the plantations, as it is stated that local causes exert a great influence in the quality of the Teas as much as the manufacture does. The expense, therefore, incurred in

---

* A little enquiry would have shown this was not true even when it was written. All Tea-planters, brokers, and all interested in Tea know now (many knew it then) that the "Assam species," *viz.*, the indigenous, makes the most valuable Tea produced.

stocking the Government plantations with the finest kinds and varieties of Tea-plants procurable in China, though great, will be amply repaid. From them superior kind of Teas are produced. This, however, I doubt, as the Teas prepared from the first imported plants have reached a perfection not surpassed by any Chinese produce."*

The above extract is a sample of the said " Records." They abound in errors and highly colored statements, which induced many to embark in Tea on unfavorable sites, and " the red book" (it is bound in a red cover) is not exactly blessed by the majority of the Himalayan planters.

---

* There must be some misprint here, for the last sentence in connection with the preceding one is unintelligible.

CHAPTER X.

TEA SEED.

Though there is a great difference in Tea-plants (see last Chapter) the seed of all is the same, and it is therefore impossible to say from what class of plants it has been gathered.

When Tea seed was very valuable (it has sold in the Tea-fever days as high as Rs. 200 per maund) it was the object of planters to grow as much as possible, and even now the price of high class Tea seed will pay well for its production.

High class plants do not give much seed, a plantation therefore with much on it should be avoided in purchasing seed.

The Tea flower (the germ of next year's seed) appears in the autumn, and the seed is ripe at the end of the following October or early November.

It takes thus one year to form.

Seed is ripe when the capsule becomes brown, and when breaking the latter the inner brown covering of the seed adheres to the seed and *not* to the capsule.

One capsule contains 1, 2, 3, and sometimes even 4 seeds.

Though the mass ripens end of October, some ripen earlier; the capsule splits and the seed falls on the ground. If, therefore, all the seed from a garden is required, it is well to send round boys all October to pick up such seeds.

When the seed is picked end of October or early November the mass are still in capsules. It should be laid in the sun for one hour daily for two or three days until most of the capsules have split. It is then shelled, and the clean seed laid on the floor of any building where it will remain dry. Sunning it *after* shelling is objectionable.

The sooner it is sown after it is shelled the better.

If for any reason it is necessary to keep it, say a fortnight or three weeks before sowing, it is best kept *towards* germinating in layers covered with dry mould. But if to be kept longer leave it on the dry floor as above, taking care it is thinly spread (not more than one seed thick if you have space) and collected together, and re-spread every day to turn it.

For transport to a distance it should be placed in coarse gunny bags only one-third filled. If these are shaken and turned daily during transit a journey of a month will not very materially injure the seed.

For any *very* long journey it is best placed in layers in boxes with thoroughly dry and fine charcoal between the layers, and sheets of paper here and there to prevent the charcoal running to the bottom.

It is scarcely necessary to consider how Tea seed can be utilized when not saleable, for seed prevents leaf, and therefore it should not be grown if there is no market for it. It will, however, make oil, but the price it would fetch for this purpose would not compensate for the diminished yield of leaf it had caused. It is also valuable as manure mixed with cattle-dung, but it would not pay to grow it for this purpose either.

My advice therefore is to allow no more seed on the garden than you require for your own use, (even the fullest gardens require some yearly) or than you can sell at a remunerative price.

If the object is to produce a considerable quantity of seed, set apart a piece of the plantation for it, and do not prune it at all. A large number will then be produced on that piece.

If the object is to grow as little seed as possible after the pruning in the cold weather, which destroys the greater part, send round boys to pick off such of the germs as remain.

TEA SEED.

If this is done, ev̶e̶r̶ so carefully, some will escape enough say to give one maund seed from 10 acres of garden, and this as a rule is enough to fill up vacancies.

The following figures regarding seed will be found useful, but remember the higher the class of plant the less durable the seed:—

Seven maunds seed, with capsules, give 4 mds. clean seed.
One maund clean seed (fresh)   = 26,000 seeds.
　　,,　　　　,,　　(ten days old) = 32,000　,,
　　,,　　　　,,　　(one month old)= 35,030　,,

Say therefore, in round numbers, that one maund Tea seed = 30,000 seeds.

With good Tea seed, sown shortly after it is picked, about 12,000 will germinate.

If you get 5,000 to germinate with seed that has come a long distance, you are lucky.

After a two months' journey more than 3,000 at the best cannot be looked for.

My experience, with seed imported into another District from Assam or Cachar, is that more than 4,500 Seedlings cannot be expected from each Maund.

CHAPTER XI.

COMPARISON BETWEEN SOWING IN NURSERIES AND IN SITU.

In the one case the seed is placed in nurseries at the close of the year, and the young plants transplanted into the garden at beginning of the following rains.

In the other the seed is (at the same time, *viz.*, close of the year, if you can get it so soon) sown at once in the plantation where the plants are intended to grow.

Each of these plans has its advocates, who don't believe in the other plan at all! The question is which is the better?

Their respective advantages may be shortly summed up as follows:—

NURSERIES.

Advantages.—The seed may be made to germinate early by watering. After it germinates the plants can be watered from time to time as they require it. Artificial shade (a great help to the germination of Tea seed) can be given. The soil can be frequently opened, and the plants in every way better tended in nurseries.

Disadvantages.—The plants lose at least three months' growth when transplanted, and may die. The transplanting necessitates labor at a time of the year it is much wanted for other work. The expense is greater than the other plan, for there are the nurseries to make and the labor of transplanting.

IN SITU.

Advantages.—The plants gain some three months in growth by not being moved. It saves labor at the busy time, *viz.*, early in the rains. It saves all the labor of transplanting, that is, it saves labor absolutely, and gives labor when, as stated, it is much required.

Disadvantages.—If the early rains (that is rain in December, January, and February) fail but few seeds germinate. In the case of a new garden the soil must be kept clean six or seven months before it would be necessary by the nursery plan. No artificial shade can be given.

It will thus be seen that the advocates of both plans have much to urge in their respective favors. Which is better?

The advocates of each plan are guided by the climate they have planted Tea in, and the truth is simply that the better plan for one place is not adapted to another. Planting *in situ* where it will succeed is by far the cheaper and better, and it will do so wherever there are certainly cold weather and spring rains. Thus (see rain table) it will succeed in Assam, Cachar, Darjeeling and perhaps the Terai below Darjeeling. It will fail in Chittagong, Dehra Dhoon, Kumaon, Kangra and Hazareebaugh. In Chittagong, for instance, a garden could never be made by planting *in situ* or as it is generally called at stake.

In this and other matters adopt your operations to the existing climate.

Where seed is planted at stake it is well not to rely entirely on it. Make a nursery also, many plants may be killed by crickets, and the vacancies can then be filled up. Again the early rain *may* fail, and thus a whole year's labor would be thrown away.

I will now describe the above two methods of sowing seed.

CHAPTER XII.

SOWING SEED IN SITU, ID EST AT STAKE.

It is named "at stake" because stakes are put along in lines to show where the Tea trees are to be, and the seed is sown at those spots.

The *modus operandi* is very simple. A month before the sowing time (which should be as soon as you can get the seed) at each stake dig a hole at least 9 inches diameter and 12 inches deep, put the soil taken out on the sides, taking care, however, if it be on a slope, to put none *above* the hole. Do not put the soil near enough to the pit, to make it likely it will be washed back. Such soil as should be washed in ought to be the new rich surface soil. For this reason the upper side of the hole should be left free on slopes. The pits are made a month beforehand to admit of this and to allow the action of the air on the open sides to improve the mould. If lucky enough to have one or two falls of rain during the month, the holes will be more or less filled up with soil, eminently calculated to instigate rapid growth. Just before sowing fill up the pit with surrounding *surface* soil. Whether to mix a little manure with it or not is a question. If it is virgin soil and rich in decayed vegetation, I say no. If not virgin soil and rather poor, yes; but it must be strictly in moderation, not more say than a man can hold in both hands, to each hole. In filling up the hole press the soil down lightly two or three times, or it will all sink later and your seeds be far too deep.

When the above is all done there is a perfect spot for the reception of the seed. The tap-root can readily descend in search of moisture, and the lateral rootlets can spread likewise. They, the latter, will not reach the outer walls of the pit for a twelve month, and will then be strong enough to force their way through.

Now sow the seed, put in say two, three, or four as the seed

SOWING SEED IN SITU, ID EST AT STAKE.   61

is good or bad six inches apart. Push them into the soft soil one inch, and put up the stake in the centre to mark the spot.

Keep the place clean till following rains, but allow only hand-weeding near the young seedlings, and occasionally open the soil with some light-hand instrument as " a koorpee" to the depth of half an inch.

If all the seeds germinate and the seedlings escape crickets and all live, at commencement of the rains leave the best and transplant the others to any vacant spot. You will succeed with some, not with others; but do not be too anxious to take up the spare ones with earth round the roots, and thus endanger the one plant left. That the seedling left be not injured is the *great* point, the others must take their chance.

Some people believe in two or even three seedlings together, and would thus advice them to be all or perhaps two left. I do not approve of the plan, except perhaps with China plants. Plant as close as you will in the lines, but give each plant its own home.

There is another mode of planting at stake, which is, I think, better than the above.

Lay the seed in alternate layers of seed and mould in beds. The seeds may be laid *close* to each other (but not *above* each other) with mould (say two inches thick) above, and then seed again. When they begin to burst, ready to shoot out their roots, examine the seeds, by taking off the soil from each layer, every three or four days. Take out those that *have* burst, and plant with the eye or root side of the seed downwards. Put all that have *not* burst back again. Repeat the operation again and again every second or third day. Be careful and take them up before the root projects, that is directly the coating has cracked.

By this means only one seed need be put at each stake, for it is certain to germinate, and seed may thus be made to go much further. Great care is, however, necessary in this operation.

CHAPTER XIII.

NURSERIES.

CHOOSE a level site with, if possible, the command of water at a higher level, anyhow with water handy. Either irrigating or hand-watering for seed beds is a necessity if vigorous and well-developed plants are to be looked for.

The soil should be of the light friable kind recommended for the Tea-plant (see " soil") and of the same *nature* as the soil of the garden, the ultimate home of the plants. This latter is all important, for seedlings will never thrive (probably not live) transplanted into a new kind of mould, particularly a poorer kind.

If possible the soil of the seed beds should be poorer than the soil of the garden, on no account richer. Taking care it is of the same *nature* as the garden soil, choose the poorest you can find. The principle is well known in England, and it applies equally in India. From poor to rich soil plants thrive, but never the other way.

For the above reason never manure seed beds.

Artificial shade for seed beds is a necessity, at least very many more seeds will germinate when it is given.

Natural shade over seed beds is *very* bad, for, *firstly*, " the drippings" are highly injurious, and, *secondly*, shade is only required till the plants are two or three inches high, after that any shade is bad, for plants, brought up to the time of transplanting in shade, are never hardy.

Seed beds, where water is handy, should not be dug deep. If so dug, and the soil is consequently loose a long way down, the tap-root will descend quickly and will be too long when transplanted. As water can be given, when it is necessary, there is no need for the tap-root to go down low in search of

NURSERIES.

moisture.* A long tap-root is generally broken in "lifting" the seedling from the bed.

Seed beds raised, as is the usual custom, above the paths that run between them, are objectionable. They part with moisture too freely. They should, on the contrary, be below the level of the paths, and there is another advantage in this, for the said paths can then be used, partly as supports for the artificial shade, and thus do away with the expense of long wooden stakes.

As the seed beds are only required until the beginning of the following rains, there is no possibility of their suffering from excessive moisture. Where they are required to remain later, of course, this plan of making the beds lower than the paths will not do.

Seed is best sown in drills, six inches apart, and each seed two, or if space can be got, even three inches from its neighbour. This facilitates each seedling being taken up later, with more or less of a ball of earth round the roots. An all-important point (see transplanting, page 76).

The length of the beds does not signify, but the breadth must not be more than five feet, so that a man on the path on either side can reach to the middle while hand-weeding or opening the soil.

After what has been said no lengthy directions for making the beds are necessary.

Cut down, burn, or carry off all jungle, and then take out all roots, whether grass or other. Now make the surface level. After this mark off the beds and paths, the latter one foot broad only, with string and pegs. Then raise the path six inches above the spots marked off for the beds. This latter

---

* In planting "at stake" (see last Chapter) the conditions are different. There the plant is in its permanent home, and the more quickly and deeper the tap-root descends the better, as the plant will then draw moisture from low down when the upper soil is dry.

must not be done by earth from the beds, but by earth from outside the intended nursery. Next dig and pulverise the soil of the beds to a depth of six or seven inches, no more, and level the surface.

All is now ready for the seed. A string, five feet long, with a small peg at either end, is given to two men who stand on the path at either side of the bed. Each man has a six-inch measure. The string is laid across the bed, beginning at one end and pegged down at either side. A drill is then made along the string about one inch deep, and this done the string is, by means of the six-inch measure on either side, removed and pegged down again in the place for the next drill. Seeds are then sown or placed along the first drill made, two to three inches apart, and the earth filled in. This is repeated again and again till the whole bed is sown.

If the character of the seed is doubtful it must be laid in thicker, but with good seed two-and-a-half to three inches is the best distance.

The sowing finished the artificial shade has to be given. Along the paths, at five feet apart, put in forked stakes two feet long, *viz.*, six inches into the path and eighteen inches above it. Connect these with one another by poles laid in the forks; now lay other, but thinner poles attached to the first poles at either end *across* and above the bed; and again across these latter, that is, along the *length* of the beds, split bamboos, and then bind the whole frame-work here and there. The said frame-work made will then be two feet above the beds, *viz.*, eighteen inches of stake support, and the six-inch raised paths. The eighteen inches of opening all round, under the frame, that is, between the frame and the path, allows the necessary air to circulate; while the expense, danger from high winds, and the objectionable entrance of the sun at the sides, all of which high artificial shade is subject to, is avoided by this low frame-work.

NURSERIES. 65

Mats are the best to cover the frame-work. In case of accidental or incendiary fire they are not so objectionable as grass, for they burn less and slower, but mats are expensive. Any coarse grass (free from seed) will answer, and it should be laid on as thin as will suffice to give shade.

The beds may be watered, if there is no rain, a fortnight after the seed is sown, and from time to time during the dry season, whenever the soil, at a depth of three or four inches, shows no moisture.

The soil should also be kept free of weeds, and after the plants are three or four inches high, the spaces between the drills should be slightly stirred every now and then.

After the seed has germinated, and the seedlings have, say four leaves on them, the artificial shade should be taken away. But it must be done gradually, taking off portions of the grass first, so that the young seedlings may by degrees be inured to the hot sun.

Though cultivation, as described, by watering and opening the soil at times is well, these should not be done much, or the seedlings will be too large when the time comes to transplant them. Large seedlings do not, as a rule, thrive as well as moderate sized ones after being transplanted.

Among the many very absurd mistakes made in the cultivation of the Tea-plant, none exceeds the ridiculous way Tea seed used to be sown in the Government plantations in the North-Western Himalayas. The seed was sown in drills, as I have advised, but in six linear inches of the drills, where it is right to put two or at most three seeds, perhaps thirty were placed! I do not exaggerate; the drill, six inches deep, was filled with them. Many and many lacs of seeds, in those days worth many thousand rupees, were thus sacrificed. Private planters in the Himalayas, taught by the Government method, once did the same. I believe the absurd practice is exploded now.

Seed cannot be sown too soon after being picked. It is ripe early in November, so the beds should be all ready by November, and if the seed has not far to come it can thus be sown early that month.

To each maund there are in round numbers 30,000 seeds, (see page 57). The numbers of plants it will take to fill an acre depends, of course, on the distances they are set apart, (see page 72), but having decided this point, also the area to be planted, and consequently the number of maunds of seeds to be sown, (see page 57), the following table will be found useful in calculating the size of nursery required.

Table showing the size of nursery required for one maund and ten maunds seed, the drills being six inches apart, and each seed three inches or two inches from its neighbour :—

Distance each seed is set apart in the drill.	Area, in sq. inches, each seed will occupy.	Area, in sq. ft., of beds, without paths, required for each md.	Area, including paths, required for each md.	Size of nursery, including the paths, to take in for 10 mds.
3 inches ...	18	3,763	4,513 sq. feet or 501 sq. yards.	100 yards by 50 yards.
2 inches ...	12	2,500	2,993 sq. feet or 332 sq. yards.	100 yards by 33 yards.

If nurseries for more than ten maunds are required then allow 100 yards to be the breadth, and for each extra ten maunds add respectively for three or two inches (see 1st col.) 50 or 33 yards to the length. Thus 50 maunds will require nurseries 100 yards by 250 yards, or 100 yards by 165 yards, according as it is decided to plant the seed three inches or two inches apart in the lines.

CHAPTER XIV.

MANURE.

An idea existed formerly, got, I believe, from stray Chinamen, who I don't think knew much about Tea in any way that manure, though it increased the yield, spoilt the flavor of Tea. The idea is opposed to all agricultural knowledge, for high cultivation, which in no case can be carried out to perfection without manure, much improves the strength and flavor of all edibles, the product of mother-earth.

My first experience of manure to the Tea-plant was obtained in the Chittagong district from a small garden close to the station, which has been for some years highly manured. I was struck with the frequency and abundance of the flushes and the strength and flavor of the Tea. My high opinion of the Tea was later borne out by the Calcutta brokers, who think very well of it and sell it at a high price.

After-experience showed me that manuring nearly doubles the yield of plants, and that so far from injuring the flavor of Tea it improves it, while it adds greatly to the strength.

I shall therefore beg the question that manure *is* an advantage. If any planter doubts let him try it, and his doubts will soon be solved.

Any manure is better than none, but I believe the best manure for the Tea-plant (always excepting night-soil and the excrements of birds which cannot be procured) is cattle manure. It is not heating like horse-dung, and may be applied in large quantities without any risk. The fresher it

K

is applied, in my opinion, the better, for it has then far more power. If mixed with any vegetable refuse, the bulk being increased, it will go farther, but I do not think it is intrinsically any the better for it.

Of chemical manures I know nothing, nor, I believe, does any one know what chemical substances are suited to the Tea-plant, but I do not doubt some of them would be suitable, and that it would pay well to manufacture such in Calcutta, and send them to Assam and Cachar, where manure is not obtainable.

All garden refuse should be regarded as manure and buried between the plants. I allude to the prunings of the bushes and the weeds at all times from the land. To carry these off the ground, as I have sometimes seen done, is simply taking off so much strength from the soil. The greener, too, all this is buried the better.

When it is considered how much is taken from the Tea-plant, it is evident the soil will be exhausted, sooner or later, if no means are adopted to repair the waste. Where manure cannot be got the waste must be made up, as far as possible, by returning all other growth to the soil. But manure *should* be got if possible, for it will double the yield of a garden.

The best way to apply it, if enough manure is procurable, is round each plant; not close to the stem (the rootlets by which the plant feeds are not there) but about a foot from it. Dig a round trench with a *kodalee*, about 9 inches wide and 6 inches deep, at the above distance from the stem, lay in the manure, and replace the soil at top. If the plants are young the trench should be narrower, shallower, and 6 inches, instead of a foot, from the stems.

If enough manure is not procurable for this (the best plan) the most must be done with what can be got, as follows:—If the plants are full grown, and there is say 4 feet between the lines, dig a trench down the centre and lay in the manure. The plants will then be manured on two sides. If the plants are young lay the manure *near* them on two sides, if possible, but failing that even on one side. The principle is to lay the manure at the distance the feeding rootlets are, and the older the plant the greater distance these are from its stem.

As to the quantity. Say for plants four years old and upwards (if younger less will be an equivalent) one maund to 20 trees is a moderate dose, one maund to 15 trees a good dose, and one maund to 10 trees highly liberal manuring, and as much as the plants can take up.

Say in round numbers each acre contains 2,500 plants (4 by 4 a usual distance gives 2,722 plants as shown at page 72), and say the manure is procurable at three annas a maund.*

The following table shows the expense of each degree of manuring, *viz.*, 10, 15 and 20 trees per maund.

It is not too much to calculate that this will add respectively 1½, 2, and 2½ maunds of Tea per acre to the yield, and I have carried this out in the table and shown the results.

I quite believe the results shown will be obtained by manuring, and I base my opinion on practice not theory.

My only experience is with cattle-manure. I know not what quantities of chemical manure would suffice, nor what the results would be.

---

* It is brought and placed between the lines, in one garden in the Chittagong district, for one to two annas a maund!

Table showing the possible cost and result of manuring with cattle-manure.

Rate of manuring.	Maunds of manure per acre at 2,600 plants per acre.	Cost of manure at 3 annas per maund. N.B.—Ans. omitted.	Probable extra yield of Tea per acre.	Value of extra yield of Tea at Rs. 50 per maund.	Profit by manuring per acre.	Deducting the probable cost of putting in the manure, the following profit is shown per acre.
	Mds.	Rs.	Mds.	Rs.	Rs.	Rs.
One md. to 10 plants...	250	47	2½	125	78	70
One md. to 15 plants...	166	31	2	100	69	62
One md. to 20 plants...	125	23	1½	75	52	46

N.B.—I have deducted Rs. 8 for the first, Rs. 7 for the second, and Rs. 6 for the third, as the probable cost of putting in the manure as it may have to be carried from the factory to the garden. If purchased after being placed between the lines, (and if manure is bought of adjacent villagers they will so place it), the cost would be less.

The above table, of course, only applies to localities where cattle-manure can be purchased at 3 annas per maund, including carriage to the factory.

The value of the extra yield of Tea is estimated at only Rs. 50 per maund in the above table, because the leaf which will give one maund of Tea is worth no more, as follows:—

	Rs.	A.	P.
Probable price obtainable for one maund or 80 ℔s. Tea in Calcutta, at 14 annas a ℔. all round, (a fair calculation, one year with the other, if it is well manufactured) ...	70	0	0
Deduct cost, manufacture, packing, transport, and broker's charges as set out in the chapter on "cost manufacture," page 160...	16	9	0
Value of leaf which will make one maund Tea	53	7	0

But I prefer estimating it at Rs. 50 only to be on the safe side.

CHAPTER XV.

DISTANCES APART TO PLANT TEA BUSHES.

When the idea existed, which it did once, that ploughs could be used to cultivate a garden between the lines, these latter, with this object, were placed unnecessarily wide apart.

All distances may be seen in different gardens, *viz.*, 6 × 6, 6 × 3, 6 × 4, 5 × 4, 5 × 5, 4 × 3, &c., &c.

The plough-idea has nowhere been found to answer, and is exploded. Still, even for hand labour to cultivate, and for facilities in picking leaf, it is necessary there should be room enough one way to pass along. Cultivation here means digging, and space enough for this must be left between the lines. Giving so much, what is then the principle that should guide us? Clearly, with a view to the largest yield obtainable, to place as many plants on the land as it will bear.

Four feet is, I think, the best distance between the lines.

It gives space enough for air to cultivate, and to pass along, even when the trees are full grown.

Where manure is obtainable, and the soil can be kept up to a rich state by yearly applications, a garden can scarcely be planted too close.

I see no objection to trees touching each other in the lines, and advise therefore 3 or 3½ feet there, the former where the soil can be periodically manured.

On considerable slopes, to prevent the wash of soil, the plants should be placed as close as possible, say 3½ between and 2 feet in the lines.

A closely planted garden will grow less weeds than a widely planted one, and will consequently be cheaper to work.

As the expenditure on a garden is in direct proportion to the area, and the yield in direct proportion to the number of plants, (always supposing there is power enough in the soil to support them), it follows that a closely planted garden *must* be very much more profitable than the reverse.

Hybrid plants grow to a larger size than China, and should therefore have more room.

The following is a useful table:—

Table showing the Plants to an Acre, and the Acres one lakh of seedlings will cover, at the distances named.

Distances in feet.	Square feet to each plant.	Plants in one acre.	The area in acres one lakh of seedlings will cover.	REMARKS.
6 by 6	36	1,210	82½	Too wide for any plants.
6 „ 5	30	1,452	69	
6½ „ 4	26	1,675	59¾	
5 „ 5	25	1,742	57½	
6 „ 4	24	1,815	55	
6 „ 3½	21	2,074	48	For Hybrids, but still I think too wide.
5 „ 4	20	2,178	45½	
6 „ 3	18	2,420	41¼	Good distances for Hybrids.
4 „ 4	16	2,722	36¾	
5 „ 3	15	2,904	34½	
4 „ 3	12	3,630	27½	China for early return.
3½ „ 3½	12¼	3,555	28	China.
3½ „ 3	10½	4,148	24	
6 „ 3¼	19½	2,233	44¾	Hybrid.
5 „ 3¼	16¼	2,726	36¾	China.
5 „ 3½	17½	2,489	40	
3½ „ 2	7	6,223	16	Best distance for China on steep slopes.

On flat land I advise—

Hybrid ... ... 4 × 4

China ... ... 3 × 3

CHAPTER XVI.

MAKING A GARDEN.

I HAVE not very much to say on this head, as most of the operations entailed are treated separately. Still a few directions on primary matters are required.

Having selected a site and made arrangements for the Tea seed required for the first year's planting, you should commence operations early in October, either by constructing the nursery, or clearing land on the proposed site of the garden, as you may decide which mode of planting, *viz.*, "nurseries" or "sowing at stake" to adopt.

If the latter, you should begin to cut the jungle somewhat earlier, but it is no use beginning to do this before the middle of September in any case, for before that the jungle would spring up again so soon, it would be labor lost.

Before you do *anything* decide how much you will cultivate the first year, and make your arrangements for seed accordingly. Here let me advise you in no case to attempt more than 100 acres. If you do 100 really well the first year you will have done *very* well. Remember you have also buildings (though few) to construct, and trying to do too much you may simply fail in all.

Previous to October you should have made yourself thoroughly acquainted with all your land, so that you can then fix with knowledge on the best sites for your buildings, nursery, and Tea plantation.

You will find much on these matters in other chapters which should be read carefully.

These respective sites having been fixed upon, and supposing you are going to plant in both ways, from nurseries and in situ, construct the nurseries as advised under that head, page 62, and also cut the jungle on the intended garden site.

MAKING A GARDEN.

There is not much to say about cutting jungle. Cut all the brushwood first near the ground, and the big trees later, so that when they fall they may lie on the underwood. In the portion you intend to plant at stakes you will not have time to cut down the big trees, and had better simply "ring" them. If this is properly done, that is if the ring is broad enough, and deep enough (less than one foot broad and 5 inches deep for large trees is not safe) they will certainly die in a twelve month, and will not give objectionable shade more than half that time. In the part to be planted "at stake" you must burn all the cut jungle end of October, and it will be well, if you have labor enough, to send men up the big trees to cut off the branches beforehand, so that they will more or less burn with the rest. Doing this, and piling up the underwood to be burnt round the base of the big trees, will cause earlier death, and diminish the objectionable shade.

Having burnt the jungle, that is as much as *will* burn, and carried off the rest from the part to be planted at stake, dig out all the small roots, and that done, dig the whole some 4 or 5 inches deep. Then stake it off with small bamboo stakes 18 inches long, showing where the Tea trees are to be (see page 72 as to the best distances) and then make your holes, and plant your seed at each stake as directed at page 60.

See the way it is recommended to stake land as regards its lay at pages 44 and 45.

You will probably not have the ground ready before the end of November, (do not attempt more than you can do up to that date), and take care and keep the seed, as directed at pages 55 and 56, until it is sown.

For the part to be planted from nurseries the following June you have plenty of time. Nowhere have I, or any one, seen large vigorous Tea-plants under trees. It is therefore evident trees are hurtful, and no more should be left in a garden than are required for the laborers to sit under occasionally, and to collect leaf under before it is taken to the Tea house. The trees that are left should be those on the sides of roads. One to every two or three acres is ample. After therefore cutting down all the low jungle, cut down all but the said few trees (it is cheaper in the end than ringing them) and then cut off, and cut up all the branches into sizes that will burn readily. Cut up the large trunks also into lengths, for all that will not burn must be carried off later. Leave all so lying until February, then choose a day with a high wind and fire it from the windward side. It may burn some days. Then collect all unburnt into heaps, and fire again and again until nothing more will burn. Now take out all roots, big and small, and when well dry, stack all these, and what was left before, and fire again and again. The land should now be tolerably clear and can be dug at once. The roads should be marked off before this, for they are better not dug.

Now stake the land at the distances determined on, and a month before the rains, or even more, if you are so far advanced, make holes for the young seedlings at each stake, precisely like those recommended for "planting at stake," page 60. Only, if possible, these should be a little larger each way than there advised, say 10 inches diameter and 15 inches deep.

Read carefully the direction as to those pits, and follow them out here. Much of the success of your planting depends on these holes.

At the first commencement of the rains transplant, as directed, under that head at pages 77, 78, 79.

Any large heavy trunks, which cannot be easily carried off the land, may be placed longways between the lines, but the less of dead timber you leave lying about the gardens the better.

L

CHAPTER XVII.

TRANSPLANTING.

IF the pits for the plants have been all prepared, as directed at pages 60 and 75, this operation is simple enough.

A fortnight or so before it commences tip all the seedlings in the nursery. Take off only the closed leaf at the head of each young plant (see a leaf diagram page 102), so that the bud at the base of the next leaf be not injured. Doing this will make the seedlings hardier and enable them earlier to recover the transplanting.

On the day you intend to take up the seedlings from any bed, if you have water enough at command, flood the bed. This as you take up each seedling will cause the soil, being moist, to adhere better to the roots.

The difference between young plants transplanted with a ball of earth round the roots, and those moved with their roots bare, is no less than three month's growth, if even it does not make the difference between life and death.

Proceed thus to insure the former. At one long end of the bed, the lowest if it is on a slope, dig close to the first row of seedlings a trench so deep that its base shall be lower than the lowest end of the tap-roots. Then with a five-pronged steel fork (this is better than a spade, <u>for it does not cut the rootlets</u>) put in between the first and second row, and pressed down with the foot to its head, force carefully so much of the row down into the trench. Then with the hand take up each seedling separately, helping the soil with a very light pressure (so light that it shall not change the lateral direction of any of the rootlets) to adhere, and place it in a low basket sloping. Do this again and again, till two baskets are full, when they will be carried, banghy fashion, to the garden.

When the first row is finished clear away the loose soil, so that a similar trench to the first shall be formed, and then proceed as above with the second row, and so on.

No further directions for lifting the seedlings out of the nurseries are required.

All is ready for their reception in the garden if the directions at page 75 have been followed out. The work now to be detailed must be done by careful men well superintended.

In the soft soil of the lately filled up pit, described at page 60, a hole is made either with the hand or a narrow kodalee, (the former, if the soil has not settled much, will suffice) large enough and deep enough to take in the seedling with all the earth attached to it. The seedling is then put in and the soil filled in and round it, which completes the operation.

The manner though in which this is done is of great consequence, four things are all important :— (1). That the tap-root shall not be turned up at the end because the hole is too shallow. (2). That any rootlets projecting outside the attached earth shall be laid in the hole, and shall preserve, when the soil is filled in, their lateral direction. (3). That the collar of the plant (the spot where the stem entered the earth in the nursery) shall be, when the pit is filled up, about one-and-a-half inch higher than the surface of the surrounding earth. (4). That in filling in the hole the soil is pressed down enough to make it unlikely to sink later, but not enough to "cake" the mould.

The following is the consequence of failure in these four points :—

1. Probably death in any case very much retarded growth. I have planted some seedlings so purposely, the

majority died; those that lived recovered very slowly, and digging them up later the tap-root was found to have gone down after all by assuming the shape of the letter S, the growth downwards being from the head of the letter.

2. Rootlets, turned away from their lateral direction, interfere with other rootlets, and though they eventually grow right if the plant lives, they retard it.

3. Fill in as you may (unless you "cake" the soil, which induces worse evils) the plant sinks a little; thus, if not placed a little high, it will eventually be too deep. If on the other hand placed too high, the rootlets and collar will be exposed, which is an evil.

4. Unless this is attended to, the plant will sink too much and the collar be buried; likewise an evil, which it takes the young seedling some time to recover.

Only first teaching and then practice will enable either European or Native to plant well. This is how it should be done.

Take the seedling in the left hand, holding it by the stem just above the collar; then take the very end of the tap-root between the second and third fingers of the right hand, and thus put it down into the hole (you thus insure the tap-root being straight). Now judge exactly the height of the collar that it be as directed. Rest the left arm then on the ground to keep the plant steady, release the tap-root, and fill up the hole about one-third, pressing the soil lightly. The plant will then be fixed, and you can employ both hands to fill up the remainder, and keep the rootlets in a lateral position. Press the soil lightly as you do so, and when all is filled up press it down a little harder round the stem of the plant.

All the transplanting should be finished as early in the rains as possible. A seedling, planted in the first fifteen days of June, is worth two planted in July, and, after the latter month, it is generally a case of seedlings and labor lost.

Days with heavy rain are not good to plant in. Those with showers or light drizzling rain are best. When there is very heavy rain the soil "cakes" much. Fine days, if the ground is wet and if more rain may soon be looked for, are good, better though if cloudy than sunny.

Where much planting though has to be done, of necessity planting must be carried on daily for, as observed, it *must* all be finished by end of July at. latest.

In case though of a sunny break in the weather, stop planting after the second day, for early rain to young transplants is a necessity.

In making a garden, too much care cannot be given to the way seedlings are placed in their homes.

CHAPTER XVIII.

CULTIVATION OF MADE GARDENS.

As manuring, which is part of this, is treated separately, we have here only to consider the best means of stirring the soil to give air to the roots of the plants, and to keep down weeds which, if allowed, injure the yield vastly.

Unless when plants are full grown and in full bearing (and not even then unless they are planted close) it is not only not necessary but a waste of labor and money to open the soil all over the garden with a view of stimulating or cultivating the plants. Much money has been wasted in this way, for instance, in a garden planted 6 by 6 or 6 by 5, and the plants, but two years old, I have seen the whole dug many times in the year. The roots of the said plants did not protrude at that age more than a foot or so, what good could they possibly derive from the extra space dug?

The soil *over* the rootlets of Tea-plants cannot be stirred too often. The oftener it is done the oftener the trees will flush, and when young the more vigorously will they grow. What is the best way to do it?

I believe simply by digging *round* each plant. I go to show why this is. I believe the best.

Putting aside the waste incurred in digging a whole garden when not necessary, the way the soil is then dug near the plants is, I think, objectionable. The ground is dug in a straight line *up to the plant*, and in doing so if the digging is deep roots are very apt to be cut. Again, when the work is task-work, the men shirk as much as possible digging close up to the stems under the branches, and thus the soil, over much of the roots, is not stirred at all. This

CULTIVATION. 81

is not easy to detect, for you must look under the branches of each tree to see how the work has been done.

In " digging round plants" the men should follow the kodalee round the tree, and the position of the blade in the same line as the roots makes any injury very unlikely. Even if tasked, as when the work is examined, it is only round the plants, it is more readily perceived if the ground has not been stirred close up to the stems.

I therefore prefer digging round plants, with the view of cultivating them, to digging the whole garden. I believe the object is better attained. That it is much cheaper is evident.

The annulus, or space to be dug round beginning 9 inches from the stem, varies with the age of the plant. Up to two years one kodalee in width will do, and after that say 2 feet.

The draw-hoe of 8 inches wide is a better tool for the above than the kodalee, especially as it is work well suited to boys, and the "draw-hoe" is a lighter tool.

Till plants from seed at stake are a year old, and till seedlings from nurseries are the same age, calculating in the latter case from the transplanting, no kodalee or even draw-hoe should come near them. The soil round for 6 inches should be slightly opened once a month or so, but it should be done with the "koorpee."

We have now discussed the cultivation of the plants. The above often done, say once a month if possible during the season, with judicious pruning and liberal manuring, constitutes high cultivation. Did weeds not grow there would be no need to do more, but weeds do grow and must not be allowed. The richer the soil the more weeds, the more manure you apply the more weeds also.

Weeds choke the plant and diminish the yield. Weeds take from the soil, and from manure when given, the strength you want for your constantly recurring flushes. If, therefore, you have a large crop of weeds you will have a small yield of Tea.

How to stop this? There is one golden rule never let them get ahead of you. This it is true argues ample labor, but unless you *have* ample labor for the area you cultivate, better let your money lie in the Bank and not grow Tea. Reduce your area until you can keep ahead of your weeds, for keep ahead you must if you wish for success.

The secret of keeping ahead of weeds is to destroy them when young, to do this again and again, as often as they come up, never allowing them to bear seed. The kodalee, an excellent digging tool, is not good for this; you want a lighter instrument, which can go over more ground and will not open the soil in the dry season to any depth. The Dutch hoe, the widest procurable in the blade, with a long lithe handle of 6 feet, is perfect for this.

With weeds at the height fit for a Dutch hoe, *viz.*, 3 or 4 inches and not numerous (which they will not be if you have "kept ahead") a man will easily do 45 square nulls, *id. est.* 720 square yards. He would not do more than 30 nulls with a kodalee.

The Dutch hoe must be well known. It is used for weeding drives and walks in England.

To conclude shortly for "hoeing and weeding" I recommend as follows:—

Dig the whole garden thrice in the year, *viz.*, spring, rains, and autumn. Bury all weeds as you dig in trenches between the lines.

In the intervals use the Dutch hoe as often as weeds appear.

Cultivate the plants by digging round them once a month if possible.

CULTIVATION.

Do all this and you will find your garden is kept clean and well cultivated, at far less cost than you incurred for cultivation when it was choked with weeds for months together, while your yield will be at the same time much increased.

If you keep your garden thus clean, and do not allow the weeds to get ahead of you, the following table shows about the cost of each cultivation operation each time you do it:—

Detail of work.	Headman at 4½ annas.	Men at 3½ annas.	Women at 3 annas.	Boys at 2½ annas.	Total Cost.			Say in Rs.
Digging the whole surface ...	¾	12	...	...	2	13	6	3
Digging round plants ...	½	...	6	5	1	13	9	2
Dutch hoeing or weeding ...	½	4	4	...	1	14	3	2

If weeds get ahead the cost in each case will be nearly double the above.

The following table, which is as near the mark as any such estimate can be, will be found useful. It will also be made use of when calculating the cost of making a garden in Chapter XXIX, page 164.

Table showing the cost per annum of keeping up at its best 100 acres of Tea from the year it is planted until the 6th year inclusive.

Year.	Rate per acre per annum.	Per 100 acres.	Remarks.
	Rs.	Rs.	
1st	50	5,000	The year the seed is sown at stake.
2nd	60	6,000	
3rd	70	7,000	
4th	80	8,000	
5th	90	9,000	
6th	100	10,000	The plants should be large plants now, but they will not be at full bearing until the 10th year.

The above rates in the case of a 300-acre garden making will include *everything* but buildings.

M

The rates are progressive because the expenditure on the following increases, or should increase yearly:—
1. Manager's pay (say every 2nd year).
2. Assistant (first entertained say 3rd year).
3. Cost and wear of tools.
4. Cost of pruning.
5. Cost of cultivation.
6. Cost of manure.
7. General expenses.

No cost for Tea manufacture is included in the above, as this is estimated for separately. See table at page 160.

Keeping up high cultivation in every way and manuring liberally a made garden in full bearing can be kept up to its highest producing powers (including the pay of the Manager, establishment and everything else) for Rs. 100 per acre per annum.

An acre of Tea may, I am aware, be kept up in a manner for Rs. 50 or so yearly, but the profit on such a plantation must be nil.

On the contrary, with the above expenditure per acre, on a good and favorably located garden, the profit will be very large. See table at page 171.

It is with Tea as with all other cultivation. It has been proved in England and in all other countries, where really high cultivation is followed out, that the higher the system followed the greater the profit. In the manner Tea has been cultivated hitherto, and is cultivated still in many places, profit can scarcely be looked for.

CHAPTER XIX.

PRUNING.

It is stated elsewhere at length, (page 101) *why* I conceive pruning to be necessary for the Tea-plant. Whether I am right or not the fact is certain that without pruning very little leaf is produced.

Pruning must be done in the cold weather when the plant is hybernating, that is to say, when the sap is down The sooner *after* the sap goes down it is done the better, for the sooner the tree will then flush in the spring.

There have been many theories about pruning Tea bushes, but none I think worth much *practically*, for the simple reason that it is impossible to prune 250,000 plants (the number in a 100-acre garden, at 2,500 to the acre*) with the care and system a gardener prunes a favorite fruit tree. The operation *must* be a coarse one, done by ignorant men, in large numbers at one time, who can in a measure be more or less taught, and the nearer they do right the better, still really careful and scientific pruning can never be carried out on a Tea plantation.

The time to do it too is very limited. It cannot be begun before the trees have done flushing, say, at the earliest, middle of November, or continued, if early flushes and a large yield next season is looked for, beyond end of January, at the latest. Thus at the most two months and a half is all the time given.

I shall confine myself therefore to giving such directions as will be practically useful.

---

* In a 500-acre garden the number is 1,250,000, which *ought* all to be pruned in two months.

The best instrument is the common "pruning knife." It cuts far cleaner than the "shears," besides which the natives very seldom use the latter well. What is called in England a "hedge-bill" is useful to trim the outside of the trees. If required it must be got from England, as I do not think it is procurable in Calcutta. Whatever instruments are used should be kept very sharp, and for this purpose, besides sharpening them every morning on the grindstone, each pruner should be provided with a small pocket "hone."

The theory, and it is correct, is in pruning, to cut near above a bud or branch, but not near enough to injure them. The cut should be quite clean and sloping upwards, so that nothing can lodge on it. This theory can be, and must be, strictly carried out in cutting the thick stems and branches, but it is quite impossible to do it with the slender branches or twigs of the tree.

Prune so as to cause lateral growth. A Tea-plant should never be allowed to exceed, say, 4 feet in height, but the wider it is the better.

Prune off all lower branches tending downwards,* for the plant should, if possible, be clean underneath to a height of say 6 inches. This clean stem high class plants have naturally; not so the China, or the Chinese caste of hybrid.

Plants should be more or less pruned out in the centre. In the following spring young wood is then formed in the heart of the tree, and it is only young wood and shoots that give leaf.

Plants exceeding 2½ feet in height at the end of the season (and all plants of any age will) may be pruned down to 20 inches, but the thick wood must be pruned down to varying heights several inches lower.

---

* The best plan with the lowest branches is to pull them off, with a sharp downward action, as then they will not grow again.

PRUNING.

Small plants must naturally be more lightly pruned.

The best plan is, I think, to have two gangs.

The first to go ahead and cut out the thick wood (here judgment is necessary, so let them be the best men) to varying heights from about 12 to 18 inches. The second gang to follow, each with a rod 20 inches long, to cut down all the light wood left to that level.

All plants, how low or how young soever they may be, must be pruned somewhat.* The lower their stature and the less their age the less pruning they require.

Of the two extremes, at least with the Tea-plant, it is probably better to over than to under-prune. The treatment of the plants, with reference to the leaf to be taken in the spring, must be a good deal regulated by the way, or rather the extent to which they have been pruned. On this point see page 101.

The cost of pruning depends on whether it is high or low, and whether the plants are large, middling, or small. The greatest cost is about Rs. 6, the least about Rs. 3 per acre.

Let all prunings be buried between the lines of plants, if possible, before the leaves have even withered. They make capital manure, but much of the virtue escapes if they are allowed to lie on the ground any time before they are buried.

---

* But not before the 3rd year.

CHAPTER XX.

WHITE-ANTS, CRICKETS, AND BLIGHT.

Both these insects are very destructive to the Tea-plants. The cricket, however, only injures it when quite young, so we will consider that little pest first.

When Tea seed germinates, and the young seedling is 2 or 3 inches high, the cricket delights to cut the stem and carry, or try to carry, the two or three green leaves attached to the upper part into its hole. Even after seedlings are planted out, if the stems are slender, it cuts them. To the young seedlings, in nurseries or planted "at stake," they often do great harm, killing in some places one-third or so.

It is much easier to prevent their ravages in nurseries than in this latter case, simply because the spot in which they must be sought and destroyed is circumscribed in the one, almost unlimited in the other.

Only one thing can be done. Employ boys (they soon get clever enough at the work) to hunt for their holes and dig them out. The holes are minute, but run down a long way. The only plan to follow them is to put in a thin pliable stick and remove the soil along it. On getting to the bottom of the stick, if it is not the bottom of the hole, you repeat the operation till you *do* get to the bottom, and there you will generally find the cricket.

Early in the morning they can be often found and caught outside their holes. The boys employed should be paid for them by the number they catch. They can be placed alive and brought to the factory in a hollow bamboo, and then killed in some merciful way.

When once a Tea-plant has got a stem, as thick as a thick pencil, no cricket can hurt it.

WHITE-ANTS, CRICKETS, AND BLIGHT.

They are much worse in some places than others, and in my experience I have found them worse on low lands.

The white-ant is a much more formidable enemy than the cricket. They *do* (as all planters know) attack and destroy living bushes.* Whether they first attack some small dead portion or not is a question, but practically it does not signify the least, for if they do they manage to find such in about one-third of the trees in a garden. Beginning with the minute dead part they kill ahead of them as they go, and will eventually, in many cases if left alone, kill the largest trees.

They have a formidable enemy in the small black ant which exist in myriads, and kills the white-ant whenever the latter is not protected by the earthen tunnels he constructs. In many places so great is the pest, did this small black ant not exist, I believe no Tea garden could stand.

From the close of the rains to the cold weather is the worse time for white-ants, and the time the planter should guard particularly against their ravages. At that time if he examines his trees closely he will very likely find white-ants on a quarter of the whole.

Digging round the plant where they are, disturbs their runs and does much good. At the same time they should be brushed off any part of the tree they have attacked, and the tree should be well shaken.

All this, however, only does temporary good, for they often are found as thick as ever on the plant a week later.

Tobacco water is beneficial, but in wet weather it is soon washed off. I have tried also a preparation, advertised by Burn and Co., for white-ants, but have only done so quite lately, and so cannot pronounce on it. I advise others to try it.

---

* A long controversy on this point lately took place in the papers.

Kerosine oil is *very* efficient. A little is put round the stem, but it is expensive. The next best thing I know is the earth-oil (petroleum) from Burmah, and this is cheap enough. It is thick but used from a bottle, it gets heated by the sun and is then quite limpid.

When white-ants are found on a tree, a little, with a small brush, is put on the part they have attacked. They are also well shaken off, and a ring of oil is placed round the stem. My experience is that they will not attack that tree again for a long time. I was at first fearful that both it and the kerosine (the one is, I believe, only a manufacture of the other) would injure the trees, but both are safe. I strongly recommend others to try it, if they doubt, on a small spot only in the first instance.

Whatever is used, or whatever is done, white-ants must not be left to work their will in the autumn. All the trees should then be examined once at least, and once again, if possible, the following spring.

Blight (a serious matter I hear in Cachar) I know but little of. I do not remember hearing anything about it when I was there, now ten years ago. It is rare in the Chittagong district, but I have seen one or two trees attacked with it. Under its influence the young leaves get covered with brown spots and shrivel. It is most destructive to the yield of a garden.

From one or two experiments made I believe pruning off all the diseased branches, and scraping back the soil for a space of 2 feet round the stem, so as almost to lay the roots bare, will be found beneficial, but I do not speak with certainty.

All the Himalayan gardens are free from these pests detailed, except that occasionally a few crickets have been seen.

CHAPTER XXI.

FILLING UP VACANCIES.

So difficult is this to do, I have heard several planters declare they would attempt it no further, but on the contrary accept the vacancies in their gardens as an unavoidable evil.

That it is difficult I too can certify. Seedlings put into vacant spots year after year die, either in the rains they are planted or the following spring. If, however, a few yards off a fresh piece of land is taken in and planted the plants live. What is the reason? It can be nothing connected with the soil, for on adjacent spots they live and die.

It puzzled me a long time, but I *believe* I can now explain it. *First*, seedlings planted in vacant spots in a garden are never *safe*. When in the rains there are many weeds in the gardens, and it is being dug, the young seedlings are not observed, are either dug up, or injured so by the soil being dug close to them, that they shortly after die. This is, I believe, the *principal* cause of the failure, and it may be in a great measure, if not entirely, obviated by putting, *first*, a high stake on either side of the seedling, and taking care they remain there all through the rains. *Secondly*, as an additional precaution, and a very necessary one, before any such land is dug, send round boys with "koorpies" to clean away the jungle round the young plants, and at the same time open the soil slightly over their roots. Doing this "cultivates" them and the plants being apparent, with the newly stirred vacant spaces round them, are seen by the diggers, and are not likely to be damaged.

The second cause of failure I attribute to the old plants on either side of the young seedling, taking to themselves all the moisture there may be in the soil during any drought. The young seedling, whose tap-root at the time is not a long one (for it is in the spring of the year following the year of planting that this occurs), is dependent for life entirely on the small amount of moisture that exists in the soil, at that insignificant depth (say 8 inches). But on two sides of the said seedling's tap-root, and in fact surrounding it, if the neighbouring Tea bushes are full grown, are the feeding rootlets of the big plants, sucking up all the moisture attainable (the necessities of *all* plants being then great), and leaving none for the poor young seedling, which consequently dies in the unequal contest.

This last evil (in climates where there is a deficiency of spring rains, and in fact more or less in all Tea localities, for in none is there as much rain as the plants require in the spring) there is no means of avoiding as long as seedlings, after transplanting, *lose time*, the effect of the transplanting, and thus fail to attain a good depth before the said dry season.

In fact unless something is devised, I believe with many, trying to fill up vacancies is a loss of time and money.

The pits to plant in, advised at page 60, should of course be made in these vacant spots, for they help much towards the early descent of the tap-root. Still they can scarcely avail sufficiently to avoid the evil, if the plant is lying inert as is generally the case for two or three months after planting. This delay being moreover in the rains, the best growing time.

If we can devise any means to avoid this delayed growth in the young seedling after it is transplanted, then the tap-root, before the drought of next spring, will have descended low enough to gather moisture for itself ; that is from a *lower* depth, than the greater number of the rootlets of the neighbouring big plants traverse. Could this be done, and if the means above detailed are resorted to, to prevent the young plants being injured when the gardens are dug, I see no reason why vacancies should not be successfully filled up. Then might be seen, what nowhere can be seen now, a Tea garden full of plants that is with *no* vacancies.

When it is considered that many gardens in all the districts have 30 or even 40 per cent. vacancies, none less than say 12 per cent., we may strike a fair average and roughly compute the vacancies in Tea gardens throughout the country at 20 per cent. In other words, the yield of Tea from India, with the same expenditure now incurred, would be one-fifth more were plantations full! That is to say, the existing gardens in India instead of producing 18 million pounds, which they do now, would produce nearer 22 millions!

I have shown how the first evil can be obviated. I *think* the following will obviate the second.

Get earthen pots made 7½ inches diameter at the head and 7½ inches deep, like the commonest flower pots, only these should be nearly as wide at the bottom as at the top. A circular hole, 2 inches diameter, must be left in the bottom. Fill these with mould of the same *nature* as the soil of the garden where the vacancies exist. Put two or three seeds in each, all near the centre, and not more than half an inch below the surface. Place these pots, so filled, near water, and beneath artificial shade, as described at page ? .

When the seeds have germinated, and the seedlings have two or three leaves, so that you can judge which is the best class of seedlings in each pot,* root out all but one, the best one. Now remove the shade gradually, water from time to time, and let the seedlings grow in the pots till the rains Having before the rains, made the holes at the vacancies as

---

* By " best class" I mean the most indigenous class.

before described, after the first fall carry the pots to the garden and place each one near a hole.

Then plant as follows. Stand the pot on the brink of the hole, having previously with a hammer broken the bottom. Then crack the sides also gently, and deposit pot and all in the hole at the proper depth. If not enough broken, the sides of the pot may now be further detached, nay even partially removed. Now fill up with earth to the top. Pieces of the pot left in the hole will do no harm ; but it, the pot, must be sufficiently broken at the bottom to allow of the free descent of the tap-root, as also enough broken at the sides to allow of the free spreading of the rootlets.

If all this has been carefully done, so that the mould in the pot shall not have been shaken free of the rootlets, the seedling will not even know it has been transplanted. Its growth will not be delayed for a day, instead of two or three months ; and by the time the dry season comes, the tap-root will have descended far enough to imbibe moisture. More than this if a little manure is mixed with the soil, filled in to the hole, the plant will be a good sized one by the end of the rains, and will give leaf the following year!

Another plan to effect the same object. Instead of pots, use coarse bamboo open wicker work baskets. The split bamboo forming the said wicker work about half an inch wide, the interstices about one quarter of an inch square. Let the diameter of the basket be the same at top and bottom, *viz.*, 9 inches. The depth of the basket 10 inches.

When the seedlings in the nursery are large enough to enable you to select a good class of plant, transplant one into each basket previously filled with soil.*. This being done, when the plants are very young, and there being *then* no difficulty in taking them up with earth attached to their

---

* Mind again this be of the same nature as the garden soil.

short tap-roots and rootlets, they will scarcely be thrown back at all. Being near water they can also be well tended. Put basket and all into the vacant hole at the beginning of the rains, and fill up as directed for the pots. The interstices will allow the feeding rootlets to pass through, besides the basket rots quickly under ground, so quickly, it cannot impede the plant.

Seed is not sown at once in the baskets as in the pots, because the baskets would not last so long. Even putting the seedling in it during (say) February, the basket, with the occasional watering necessary, will, more or less, have rotted before it is put into the hole.

I have concluded a contract for ten thousand pots and five thousand baskets at half an anna each for both kinds. Two pice, to ensure the filling up of a vacancy, is not a large outlay.

Since writing the above I have had experience of both the above plans. The pot system is far the better, and answers very well.* I am now trying to improve this still further by making the pots a little larger, and placing a thin inner lining of tin inside each about half an inch from the sides. This space is then first filled with sand, then the pot is filled with mould, and the tin pulled out. The same tin will therefore do for any number of pots. The seed is then put in.

I think by this plan if when about to plant the mould in the pot is well wetted, that it, with the seedling, can be turned out whole in one piece, and then put in the hole *without* the pot.

The same pots would then answer year after year, and the expense would be quite nominal.

If well done, the seedling in this, as in the former case, would not even *know* it had been transplanted.

---
* The baskets are too frail—Being often wetted it falls to pieces before the planting time.

CHAPTER XXII.

FLUSHING AND NUMBER OF FLUSHES.

The Tea plant is said to flush when it throws out new shoots and leaves. The young leaves thus produced are the only ones fit to make Tea, and the yield of a plantation depends therefore entirely on the frequency and abundance of the flushes.

The way a flush is formed is fully explained under the head of "leaf picking," (page 102.)

The number of flushes in different plantations varies enormously, owing, *first*, to climate; *secondly*, to soil; *thirdly*, to the pruning adopted; *fourthly*, to the degree of cultivation given; and *fifthly*, though not least, to the presence or absence of manure.

How to secure all these advantages to their fullest extent is shown under those heads, and we have here only to consider what is a low, a medium, and a high rate of flushing per season.

In doing this we must speak of elevated (as Himalayan) gardens separately. The cool climate of heights makes it impossible for Tea to flush there as on the plains.

Speaking generally of elevated gardens (the higher they are the shorter the period, and *vice versâ*) seven months may be considered as the average producing period, *viz.*, from beginning of April to end of October, and during that time 12 to 15 flushes may be obtained, which, I believe, with high cultivation and liberal manuring, can be increased to 18.

In all localities, with favorable Tea climates, the plants flush both for a longer period and oftener, speaking generally also, in this case, of the four best localities, *viz.*, Assam, Cachar, Chittagong, and the Terai below Darjeeling (for even in these districts many advantages exist on one garden which do not in another) the following is an approximation to the flushing periods:—

Upper Assam.—February 25th to November 15th.
Lower Assam.—February 20th to November 20th.
Cachar.—February 20th to November 20th.
Chittagong.—March 10th to December 20th.
Terai below Darjeeling.—March 1st to November 20th.

The opening period is a little late in Upper Assam on account of the cold, and closes a little earlier for the same reason.

Lower Assam and Cachar are much alike.

The opening in Chittagong is later than in the two just mentioned from want of early rains, but the season continues longer on account of the low latitude and consequent deferred cold weather.

Roughly, then, rather more than nine months may be assumed as the flushing period for these districts. The next point is how *often* do gardens in these localities flush in that time.

Not very many planters can say, certainly, how often their gardens have flushed in a season, because they are picked so irregularly, and no account of the different flushes kept. Enquiring on this point, when I was in Cachar some ten years ago, 9 to 24 were the minimum and maximum numbers given me at different gardens, showing how little was really known about it.

Such knowledge as I have on the subject is mostly derived from carefully kept records of my own garden in the Chittagong district. The plantation is all worked in sections, in the way described previously, and the dates given in the table below are the days each flush was finished (that is, the picking was finished) during the seasons 1869 and 1870; 1869 being carried up to the end of the season, 1870 up to the date I wrote the first edition of this essay.

In the table it will be observed there is a great difference between the two years. The section for which the dates are given was planted from seed beds in the month of June 1866. In 1869 it was therefore only three years old. This will partly account for the first flush occurring a month earlier in 1870, as it was then a year older; but fortunate early rains in 1870 had also much to do with it.

In 1869 there was no flush between March 22nd and May 6th, a period of 44 days; and in 1870, none between February 22nd and March 30th, a period of 35 days, a very long time in both cases, which is entirely accounted for by the dry weather prevailing at Chittagong in the spring (see under head of Climate) for in Cachar or Assam two or three flushes would have occurred in that time.

There were 19 flushes in all in 1869, and 22 in all in 1870, up to the time I wrote, so there were probably 27 that year.

In the table I give the intervals between each flush. It shows an average of 14 days in 1869 to 10 days in 1870, the difference is due to the increased age of the plants, and the liberal manuring given in the cold weather 1869-70.

Flushes.	1869. Dates.		Interval in days.	1870. Dates.		Interval in days.
1	March	22	......	February	22	......
2	May	6	44	March	30	35
3	,,	29	23	April	13	10
4	June	11	12	,,	25	12
5	,,	23	12	May	5	9
6	July	5	11	,,	14	9
7	,,	17	12	,,	25	11
8	,,	31	14	June	4	9
9	August	10	9	,,	12	8
10	,,	21	11	,,	22	10
11	Sept.	2	11	July	1	8
12	,,	12	10	,,	8	7
13	,,	25	13	,,	16	8
14	October	9	13	,,	25	9
15	,,	22	13	August	2	7
16	Nov.	2	10	,,	11	9
17	,,	11	9	,,	21	10
18	,,	19	8	,,	29	8
19	Dec.	4	14	Sept.	7	8
20	......		......	,,	18	11
21	......		......	,,	27	9
22	......		......	October	5	7
Average intervals between Flushes.			Nearly 14 days.	......		Very little over 10 days.

Such a result as is shown for 1870, and the probable result of 27 flushes to the end of that season, could not be obtained without high cultivation and liberal manuring. The land in question had been manured every year since it was planted, but an extra dose was given in the cold weather of 1869-70. The ground was therefore very rich.

I think therefore 25 flushes in the season may be looked for on gardens in good Tea climates, when high cultivation and liberal manuring is resorted to. Where manure cannot be obtained, I think, even if in other respects the land is highly cultivated, more than 22 flushes will not be obtained. Where neither manure nor high cultivation are given, above 18 flushes will not be got.

It seems to be a general idea with planters (see diagram, page 102,) that when a flush is picked the succeeding flush, at an interval of say ten days, are shoots from the axis of the leaf down to which the previous flush was picked. Thus in the diagram supposing the shoot to be picked down to the black line above 2. The idea is the next flush will be a shoot springing from the same place, *viz.*, the axis of leaf *d*. But it is *not* so. In the above case it will take a whole month, after the said shoot has been picked, before the new shoot from the base of the leaf *d* is ready to take, probably six weeks in Himalayan gardens.

'Tis true the flushes follow at about ten days from each other, but they are *other* shoots. The replacement of the shoot taken is a whole month in developing. I have carefully watched this and am sure I am right.

With similar treatment gardens in Cachar and Assam would probably give two or three more flushes in the season than Chittagong, because there the spring rains are much more abundant; and I am very certain that, if the day ever comes that manure in large quantities is procurable in those districts and is applied, the yield on those gardens will be very large.

The difference between very small and very large profits is represented by 18 and 25 flushes, so I strongly advise all planters to cultivate highly and to get all the manure they possibly can. If even procured at a high figure it (the manure) will pay hand over hand.

CHAPTER XXIII.

LEAF-PICKING.

THE first consideration is how to get the largest quantity of leaf without injuring the trees.

To a certain extent, it is true, that the more a Tea bush is pruned and picked the more it will yield. It appears as if nature were always trying to repair the violence done to the tree by giving new mouths or leaves to breathe with in place of those taken away. I may exemplify my meaning another way. A Tea bush which has as many leaves on it *as it requires* will throw out tardily new shoots, and their number will be small. In other words, a plant which is not pruned and from which the young leaves are not taken grows gradually large and bushy, and then gives up flushing altogether. It has all the leaves it *requires,* and it has no necessity to throw out more.

If, however, nature is too much tried, that is if too much violence is done to her, she sulks and will exert herself no more. Up to this point, therefore, it is well to urge her. How can we know when we have reached it?

Only general rules can be laid down. Experience is the great *desideratum* on this, and many other subjects connected with Tea.

If the plant can always be kept in such a state that the foliage, without being *very much so,* is still less than nature requires, I conceive the object will be attained.

The greatest violence is done to the plants when it is pruned, and reason would seem to argue that when this violence is repairing, that is when the first shoots in the spring show themselves, and until new mouths (or leaves) in sufficient quantities exist, until then but little leaf should be picked.

Fortunately, moreover, while in the interests of the plant this is the best plan, it also is the mode by which the largest yield of leaf will be secured in the season. I go to show this.

The ordinary size of a good full-grown Tea-plant, at the end of the season, is say 3½ or 4 feet high, and 5 feet diameter. It is pruned down say to a height of 2, with a diameter of 3 feet. It is then little more than wooden stems and branches, and to any one ignorant of the *modus operandi* in Tea gardens, it would appear as if a plantation so pruned had been ruined. The tree remains so during all its hybernating period, that is during the time it is resting and the sap is down, (this period is longer or shorter, as the climate is a warm or cold one, and it is always during the coldest season) but on the return of spring new shoots start out from the woody stems and branches, in the following way:—At the axis or base of each leaf, is a bud, the germ of future branches, these develop little by little, until a new shoot is formed of, say 5 or 6 leaves, with a closed bud at top. Then if it be not picked the said bud at top hardens. At the axis or base of each of the said 5 or 6 leaves are other buds, and the next step is for one, two, or three, of these to develop in the same way and form new shoots. The original shoot grows thicker and higher until it becomes a wooden branch or stem. The same process, in their turn, is repeated with the new shoots. A diagram will make my

meaning clear. We here have a shoot, fully developed of six leaves, counting the close leaf a at top as one, viz., the leaves a, b, c, d, e, f. The shoot has started and developed from what was originally a bud at K, at the axis or base of the leaf H. In the same way as formerly at K a bud existed, which has now formed the complete shoot or flush K a, so at the base of the leaves c, d, e, f, exist buds 1, 2, 3, 4, from which later new shoots would spring. These again would all have buds at the base of the leaves, destined to form further shoots, which again would be the parents of others, and so on to the end of the season, or until the tree is pruned.

It will readily be seen the increase is tremendous. It is only limited by the power of the soil to fling out new shoots, and the necessities of the plant, for as I have explained, when as much foliage exists as the plant requires, but few new shoots are produced.

Now supposing the shoot in the diagram to be (with perhaps another not shown at L) the first on the branch I.I.I. in the spring (the said branch having been cut off or pruned at the upper I). It is then evident the said shoot is destined to be the parent and producer of all the very numerous branches and innumerable shoots into which the plant will extend in that direction. It is in other words the goose which will lay all the future eggs. If, eager to begin Tea making early, the planter nips it off, the extension on that part of the tree is thrown back many weeks. It may be taken off at 1, 2, or 3 (the back lines drawn show the proper way to pick leaf), the least damage will be done if it is taken off at 1, the most at 3.

The said shoot K a is the first effort of nature to repair the violence done to the tree by pruning. It is the germ of many other branches and shoots, and it ought *never* to be taken. I have, I hope, made so much plain.

There is, however, another consideration. Any shoot, left

to fully develop and harden, does not throw out new shoots from the existing buds 1, 2, 3, 4, so quickly as one checked in its upward growth by nipping off its head. For instance, supposing the shoot under consideration *not* to be the first of the season, but on the contrary to be a shoot, when the plant has developed sufficiently to make picking safe. If taken off at 2, then the new growth from 2, 3, 4 will be much quicker than it would be had the whole shoot been left intact.

Our object then with *first* shoots should be to secure this advantage without destroying any buds, and this we can do by taking off simply the closed leaf at the top *(a)*. This must be done so as not to injure the bud at the base of the second leaf *b* (I have not numbered it, for there is no room in the diagram to do so), and we shall thus leave all the buds on the shoot intact.

Again here the interests of the plant, and profit to the planter, go hand in hand. The closed bud *a* in this case will be found very valuable. I go to show this.

The value of Tea is increased when it shows "Pekoe tips." Only the leaves *a b* make these. They are covered with a fine silky whitish down, and if manufactured in a particular way make literally white or very pale yellow Tea,* which mixed with ordinary black Tea show as "Pekoe tips." In ordinary leaf-picking these two leaves are taken with all the others, but unfortunately, when manufactured with them, they lose this white or pale yellow color, and come out as black as all the other Tea.

As the season goes on, though this is less and less the case, till towards the end, nearly all the *a b* leaves show orange colored in the manufactured Tea. They are then never however *white* (the best color) as they can be made when treated separately. No means have yet been devised to separate

---

* I mean manufactured Tea. The infusion is called liquor.

them *before* manufacture from the other leaf, and though sometimes picked separate, the plan has serious objections (see page 106). In the case, however, of the first 2 or 3 flushes the welfare of the plants demands that no more should be taken, and though the quantity obtained will be small, it will, if carefully manufactured so as to make " white Pekoe tips," add one or two annas a ℔ to the value, when mixed with it, of one hundred times its own weight of black Tea!

More will be found under this head in the Tea manufacturing part. I now beg the question that the said downy leaves taken alone are very valuable.

In detailing the mode of picking I advocate, it would be tedious to go minutely into the reasons for each and every thing. I have said enough to explain a good deal, but will add any thing of importance. Of the latter are the following.

Tea can be made of the young succulent leaves only. The younger and more succulent the leaf the better Tea it makes. Thus a will make more valuable Tea than b, b than c, and so on; e is the lowest leaf fit to make Tea from, for though a very coarse kind can be made from f, it does not pay to take it. The stalk also makes good Tea, as far as it is really succulent, that is down to the black line just above 2.

The leaves are named as follows from the Teas it is supposed they would make:—

a.—Flowery Pekoe.
b.—Orange Pekoe.
c.—Pekoe.
d.—Souchong, 1st.
e.— ,, 2nd.
f.—Congou.

Mixed together ... $\begin{cases} a, b, c\text{—Pekoe.} \\ a, b, c, d, e\text{—Pekoe Souchong.} \end{cases}$

If there be another leaf below *f*, and it be taken, it is named, and would make Bohea.

Each of these leaves was at first a flowery Pekoe leaf (*a*), it then became a *b*, then *c*, and so on.

That is to say, as the shoot developed, and a new flowery Pekoe leaf was born, each of the leaves below assumed the next lowest grade.

Could the leaves fit to make each kind of Tea, it is proposed to make, be picked and kept separate, and each be manufactured in the way most suitable to its age, and the Tea to be produced, the very best of every kind could easily be manufactured. But this cannot be; the price of Tea will not allow it, and the labour to do it would moreover fail. It has been attempted again and again to do it, partly to the extent of taking the Pekoe leaves *a*, *b*, *c*, separate from the others (for the manufacture best suited to these upper leaves is not suited to the lower), but it has been as often abandoned, and I doubt if it is now practised anywhere. I am sure it will never pay to do it.

Picking leaf is a coarse operation. It is performed by 80 or 100 women and children together, and it is impossible to follow each, and see it is done the best way. They must be taught, checked, and punished if they do wrong, and then it will be done more or less right; but perfection is not attainable.

I advise the following plan in picking. Please refer to the diagram :—

If the garden has been severely pruned (as it ought to be) take only the bud *a* for *two* flushes; then for *two* more nip the stalk above 1, taking the upper part of leaf *c*, as shown (done with one motion of the fingers). Then from the 5th flush take off the shoot at the line above 2, and by a separate motion of the fingers take off the part of leaf *e*, where the black line is drawn. By this plan, when the rains begin, the trees will show a large picking surface, for plenty of buds will have been preserved for new growth. After the month of August you may pick lower if you like, as you cannot hurt the trees. For instance, you may nip the stalk and upper part of leaf *e* together, and separately the upper part of *f*.

The principle, however, of picking is to leave the bud at the axis of the leaf down to which you pick intact.

Some planters pick all through the season at the line above 1, and take the *d* and perhaps the *e* leaf separately. I do not like the plan, for though it will make strong Teas, the yield will be small. Moreover, the plants will form so much foliage; they will not flush well, and again they will grow so high that boys who pick will not readily reach the top.

Shortly the principle I advocate is to prune severely, so that the plant in self-defence *must* throw out many new shoots. To be sparing and tender with these until the violence done to the tree is in a measure, but not quite, repaired; then, till September, to pick so much that the wants of the plant in foliage is never quite attained, and after September to take all you can get.

I believe this principle (for the detailed directions given may be varied, as for instance when trees have *not* been heavily pruned) will give the largest yield of leaf, and will certainly not injure the plants.

P

CHAPTER XXIV.

'MANUFACTURE. MECHANICAL CONTRIVANCES.

To manufacture your leaf into good Tea is certainly one of the first conditions for success. It will avail little to have a good productive garden if you make inferior Tea. The difference of price between well and ill-manufactured Tea is great, say 4 as. or 6 d. a ℔, and this alone will, during a season, represent a large profit or none.

Fortunately for Tea enterprise, the more manufacture is studied the more does it appear, that to make good Tea is a very simple process. The many operations or processes, formerly considered necessary, are now much reduced on all gardens. As there was then, that is formerly, so there is now, no *one* routine recognized by all, or even by the majority; still simplicity in manufacture is more and more making its way everywhere, and as the real fact is that to make the best Tea, but very few, and very simple processes are necessary, it is only a question of time, ere the fact shall be universally recognised and followed out.

For instance, panning the "roll"*? was formerly universally practised. Some panned once, some twice, some even three times! But, to-day, pans are not used in most gardens at all!! Other processes, or rather in most cases the repetition of them have been also either discarded or abridged. But a short statement of manufacture in old days,

---

* In describing manufacture I shall call the leaf brought in "leaf" until it enters on the rolling process. From that time until the drying over charcoal is concluded "Roll," and after that "Tea."

MANUFACTURE. 109

and the simplest mode of manufacture now will best illustrate my meaning:—

Days.	Number of operations.	Detail.	Days.	Number of operations.	Detail.
		One and a common old plan.			One plan to-day by which the best Tea can be made.
1st	1	Withering.	1st	1	Withering.
	2	1st Rolling.		2	Rolling.
	3	2nd ,,	2nd	3	Fermenting.
	4	Fermenting.		4	Sunning (if sun.)
	5	1st Panning.		5	Firing (Dholing.)
2nd	6	3rd Rolling.			
	7	2nd Panning.			
	8	4th Rolling.			
	9	Sunning.			
	10	1st Firing (Dholing.)			
	11	Cooling and crisping.			
3rd	12	2nd Firing (Dholing)			
3	12	Total days and operations.	2	5	Total days and operations.

So much for simplicity, and I affirm that no more than the five operations detailed are necessary. I shall try to show this further on.

In studying Tea manufacture I first tried, in order to get reliable data to go on, to ascertain the effect of each and every operation, and not only that but the effect on the made Tea of each operation exaggerated and diminished. It would be tedious, and of no use, to set out in detail all the experiments I conducted, the results only I will try and give.

MANUFACTURE.

I began at the beginning. Why wither at all? I made Tea (following out in each case all the other processes detailed in the old plan) of 1st, totally unwithered leaves; 2nd, of leaves but little withered; 3rd, of leaves medium withered; and 4th, of leaves over-withered.

I arrived at the following results:—Unwithered or under-withered leaves break in the rolling and give out large quantities of a light green colored juice during the same process. The Tea is much broken and of a reddish grey color. The liquor is very pale in color, cloudy, weak, soft, and tasteless.

Over-withered leaf on the other hand takes a good twist in the rolling, gives out but little juice which is of a thick kind, and of reddish yellow color. The Tea is well twisted, "chubby" in appearance, and blacker than ordinary. The liquor of an ordinary depth of color, clear with a mawkish taste.

The medium withered leaves made good Tea, but I found the withering should be rather in excess of what is generally done to ensure strength. I will show later to what extent I think leaf should be withered.

The next point was rolling. I knew some planters rolled the leaf hard, others lightly. That is, some rolled with force till much juice was expressed, others with a light hand, allowing little or no juice to be pressed out. Which was the better?

After many experiments I arrived at the following:—Hard rolling gives darker colored and stronger liquor than light rolling. Hard rolling destroys Pekoe tips,* inasmuch as the juice expressed stains them black.

Light rolled Tea has therefore many more Pekoe tips than hard rolled.

Hard rolled Tea is somewhat blacker than light rolled.

---

* Pekoe tips are the whitish or orange colored ends that may be seen in Pekoe Tea. See page 114.

In all, therefore, but the point of Pekoe tips hard rolling is better.

The next question was what is the advantage of repeated rollings? I rolled twice, panning once between, *vide* old plan, and found the Tea as well made, and as strong as others rolled three or four times. I then decided to roll *no more* than twice. The second time was, I *then* thought necessary, as I found the leaf of the roll opened in the pan, and a second rolling was requisite to twist it again.

But what did panning do? I heard pans had been discontinued in some gardens. In what way was then panning an advantage? I made Tea, fermenting it between the two rollings, but *not* panning it, and it was equally good. I tried again and again, but never could detect that panning caused any difference to either the Tea, the liquor, or the out-turn.* In short, though I never found panning did any harm, I equally found it never did any good. Its use is, in fact, simply barren of *all* results.

I therefore dispensed with it. Having done so, why roll the second time at all? I experimented and found the second rolling as barren of results as the panning.

I had now got rid of operations 3, 5, 6, 7, and 8 in the old plan. The next was No. 9 "sunning." I made Tea with, and without it, and found as follows:—

Sunning between the fermenting and firing processes has no effect whatever on the liquor or the out-turn, but it makes the Tea rather blacker, and as it drives of and much of the moisture in the roll, the firing process after it is shorter and does not consume so much charcoal. What little effect therefore it has is good (for if not continued too long, it

---

* The out-turn are the Tea leaves after infusion.

does not make the Tea too black) and it is economical. I therefore decidedon retaining it.*

Next came the operations 10, 11, and 12, *viz.*, "first firing," "cooling and crisping, and second firing." Where these are done (and they are done in some gardens now) the usual thing is to *half-fire* the roll the same afternoon and evening it is made, then allow it to "cool and crisp" all night, and finish the firing next day. I tried this plan, and also the plan I have now adopted, of doing the whole firing at one time the same evening. I tried the experiment again and again, and always found the Tea, the liquor, and the out-turn was the same in both cases. In short that the three operations did no more, and no less than the one. As the three entail extra labour and extra expense in charcoal I abandoned them.

I thus reduced the twelve operations detailed to five, and naturally by so doing much decreased the cost of manufacturing Tea. I in no way lay claim to having devised this simplicity myself. Part had been done by others before I ever turned my attention to it, and I have done no more than help with many to make the manufacture of Tea a simple process.

I was now convinced that, (though I had still much to learn regarding the said five processes), success was comprised therein, and that to multiply them could not avail.

The next consideration is—What are the qualities desired in Tea to enable it to command a good price at the public auctions, either in Calcutta or London ? The brokers in these cases judge of the Tea first, value it, and give their report and valuation to intending purchasers and sellers. From what appearances and qualities do they judge ?

---

* At the end of the season, however, sunning has more than the above effect. It then makes the Tea "chubby" in form, of a reddish color, and improves the strength of the liquor.

They judge from three things, *first*, the Tea; *secondly*, the liquor; *thirdly*, the out-turn.

The Tea.—The color should be black, but not a dead black, rather a greyish black with a gloss on it. No red leaf should be mixed with it, it should be all one color. The Tea should be regular; that is, each leaf should be about the same length, and should have a uniform close twist, in all but " broken Teas." (These latter are called " broken," *because* the leaf is more or less open and broken). The Tea should also be regular of *its kind*, that is, if Pekoe all Pekoe, if Congou all Congou; for any stray leaves in a Tea of another kind, if even of a *better* kind or class, will reduce its value. In the higher class of Teas, *viz.*, Pekoes and Broken Pekoes, the more Pekoe tips that are present the higher, in consequence, will its price be.

The Liquor.—In taste this should be strong, rasping, and pungent, with, in the case of Pekoes, a " Pekoe flavor." There are other words used in the trade to particularize certain tastes, but the words themselves would teach nothing. Tea-tasting cannot be learnt from books. *If* the liquor is well flavored; as a rule, the darker it is in the cup the better. But to judge of Teas by the color of the liquor alone is impossible, for some high class Teas have naturally a very pale liquor.

The Out-turn.—A good out-turn is generally indicative of a good Tea. It should be all, or nearly all, one color. No black (burnt) leaves should appear in it. A greenish tinge in some of the leaves, is not objectionable, and is generally indicative of pungent liquor, but the prevailing color should be that of a bright new penny.

Every planter should be more or less of a Tea-taster, and should taste his Teas daily. After a time (particularly if he gets other Teas to taste against his own) he will learn to recognize, at all events, a good as against a bad Tea, a strong as against a weak Tea, &c. No Tea should be put away with the rest until it has been tasted. It may be burnt or have other defects, not apparent till infused, and one day's bad Tea will bring down considerably the value of a whole bin of good Tea.

The fancy, amongst brokers and dealers, for "Pekoe tips" in all Pekoe Teas, constitutes the *one* great difficulty in Tea manufacture. If the leaves which give "Pekoe tips" (see page 104) are separated from the other leaves, and manufactured separately and differently, that is rolled *very* little and *very* lightly, not allowed to ferment at all, but sunned at once after rolling, and if there is sun enough finished in the sun, otherwise by a very (light and gradual heat) best placed *above* the drawers in the (Dhole-house) if this is done, I say, these will come out perfect "Pekoe tips" of a white color, which is the best.

If *not* separated from the other leaf, but manufactured with it, the sap from the other leaves, expressed in the rolling, stains these said leaves, which are covered with a fine white silky down, and makes them black like all the rest of the Tea; the whole of which is then valued lower, *because* there are no "Pekoe tips."

Now in the latter case " the Pekoe tips" are there all the same, only they don't *show*. The Tea is really just as good, in fact a shade *better* with black than with white or orange tips,* but it does not sell so well, and as we cannot argue the brokers, or dealers into a rational view of the case, we must humour their fancy (they are virtually our masters) and give them the Pekoe tips,—*if we can.*

How are we to do it? The plan of picking these small leaves separately, in order to manufacture them separately, does not answer. It is too expensive; it diminishes the

---

* It is better because "the tips" having been hard-rolled give stronger liquor.

yield of a garden, and labor for it fails. All this is shown at page 106. Is there any other way?

It may be done during some periods of the season when there is not leaf enough on the garden to employ all the leaf-pickers, by setting a number of them to separate the said two leaves from the others *after* the whole leaf is brought to the factory. This is expensive, but it pays when there is labor to do it, for then the Teas can be made very showy and rich with white Pekoe tips.

An ingenious planter, a Mr. McMeekin in Cachar, invented a rolling table with the object of separating the said leaves. It is constructed of battens, and while rolling the leaf on it, many of the small leaves fall through. The said table is now well known in Cachar, and is in use in several gardens. I have tried it and find that it in a great measure answers its object, but the objection to it is that the leaf *must* be rolled lightly, and lightly rolled leaf, as observed, does not make strong Tea.

The Pekoe tips may be, in a great measure, preserved by rolling all the leaf lightly on a common table. But then again the Tea is weak, and the plan will not give so many Pekoe tips as McMeekin's table.

In short, in the present state of our knowledge, except by the hand process, (a tedious and expensive one for separating the leaf), strong Teas and Pekoe tips are incompatible. The difficulty is just where it was, and will so remain until dealers give up asking for Pekoe tips (not a likely thing) or till a machine is invented, to separate quickly and cheaply, the two said small leaves from the others *after* they have been all picked together. That such a machine is possible I am certain, and the inventor would confer a boon on the Tea interest far beyond the inventor of any other machine, for all the other processes *can* be done by hand without much expense, this cannot.

I may here notice such machines and contrivances as exist for cheapening the manufacture of Tea, or rather such as I know of.

Rolling machines have for their object the doing away with hand labor entirely for rolling the leaf. Kinmond's rolling machine is first on the list, for it is the best yet invented.*

Kinmond's consists of two circular wooden discs, the upper one moving on the lower, which is stationary, with an eccentric motion. The adjacent faces of the said discs are made rough by steps in the wood, cut in lines diverging from the centre to the circumference, and over these rough faces is nailed coarse canvas.

The leaf is placed between the discs and rolled by the motion described. The lower discs is arranged by means of weights running over pulleys, so that it shall press against the upper with any force desired.

The motive power, as designed by the inventor, is either manual, animal, or steam.

Mr. Kinmond showed me this machine, just after he had invented it, at the Assam Company's Plantations in Assam some seven years back, and I have since seen it working by manual and steam power. With the former it is quite useless, for by no arrangement can sufficient or regular force enough be applied. With the latter it does very well, and on a large garden which will render the outlay for the machine, and engine justifiable (the former is, for such a simple machine very expensive) it may probably eventually prove an economy.

Not having seen it under animal power, I can give no positive opinion as to how it would answer, but I see no reason why it should not do well. I believe wind or water power might, on suitable sites, be easily applied to it, and they would certainly be the cheapest of any.

---

* It was the best, but is superseded by a new rolling machine, Jackson's, I have seen quite lately. See the Addenda at the end.

Another rolling machine was invented by a Mr. Gibbon, and a good deal used in Cachar. I have never seen it.

Kinmond's is, I believe, the best rolling-machine yet invented (though it is fair to state I know no other except by report), but I do not believe in any Tea rolling-machine superseding *entirely* the necessity of hand rolling*. A rolling-machine may be, and is very useful, to roll the leaves partly, that is, to break the cells, and bring the leaf into that soft mashy state that very little hand labor will finish it. No rolling-machine yet invented can, I think, do more than this, and it is, I think, doubtful, if any will ever be invented that will do more. Machines do not give the nice final twist which is obtained by the hand. I was told lately that most of the gardens in Cachar that had machines had dropped them and gone back to hand-rolling. I cannot help thinking this is a mistake. They should use both, the hand-rolling for the final part alone. Very few rolling-men would then suffice with aid of the machine to manufacture a large quantity of leaf.

I only know of one other Tea rolling-machine which is Nelson's. It does not profess to do more than *prepare* the green leaf for rolling, which, as stated above, is, I think, all that any machine will ever do. I have never seen it working, but it appears simple, being nothing more than a mangle. The leaf is placed in bags, and then compressed under rollers, attached to a box, weighted with stones. The prospectus states, it will prepare 80tbs. green leaf in fifteen minutes, and that one man can then finish as much of such prepared leaf in three minutes, as would occupy him twelve minutes if the same had not been prepared. I see nothing unlikely in this. The machine, though inferior to Kinmond's in its arrangement, *ought* to be cheap enough to bring it within the reach of all.†

---

* I had not seen Jackson's machine when I thought as above.
† Unfortunately it is not. It is advertised at Rs. 300, with a yearly royalty of Rs. 50 the first year and 20 after. The royalty should be dropped, and the machine sold for Rs. 150, which would give the inventor a good profit.

I have already spoken of one of McMeekin's inventions. His chest-of-drawers for firing Tea is, I think, superior to his batten table. It is now so well known, and in such general use, that I shall describe it very shortly. It is nothing more than a low chest-of-drawers, or trays fitted in a frame one above the other, the bottom of each tray being fine iron wire, so that the heat from the charcoal, in the masonry receptacle over which it is placed, ascends through all the drawers and thus dries or fires a large quantity of "roll" at the same time. By the old plan, a single wicker sieve was inserted inside a bamboo frame, called a "dhole," which was placed over a charcoal fire made in a hole in the ground. On the sieve the roll was placed, and all the heat after passing through this *one* sieve was wasted. Mr. McMeekin's idea was to economize this heat by passing it through several drawers.

Most planters use these drawers, and there is no doubt in the space saved, and the economy of heat, it is a great step in advance over the old barbarous method, where not only was the heat wasted after passing through *one* sieve, but a great deal was lost through the basket work of the "dhole" itself.

Still I do not advocate four, still less five drawers one above the other. I think the steam ascending from the lower drawers must, more or less, injure the roll in the upper ones. I confine myself to two, and even then in the top tray leave a small circular space vacant by which the steam from the lower drawer can escape. I utilize the heat that escapes, partially, by placing "dhallas" in tiers above, with roll in them. These are supported by iron rods let into the wall and are useful, not only for partly drying the roll, but also for withering leaf when there is no sun.

Some planters have proposed to do away with charcoal altogether under McMeekin's drawers, supplying its place by hot-air. The first point in considering this invention is the question whether the fumes of charcoal, as some assert, *are* necessary to make good Tea. If they are *not* necessary (that is, if they produce no chemical effect on the Tea, and therefore heat from wood, devoid of smoke would do as well) there can be no doubt such heat would be cheaper, and more under command, by this or some other plan. Are then the fumes of charcoal necessary?

I do not know that any one can answer the query. I certainly cannot, for I have never made Tea with any other agent than charcoal, and I have never met with more than one planter who had. *He* said the Tea was not good. Still it would, I think, require very careful and prolonged experiments to establish the fact either way. Speaking theoretically, as it *appears*, the only effect of charcoal is to drive all the moisture out of the roll and thus make it Tea, I cannot but believe other heat would do as well. It is however a question that only experience can solve.

I have now (four years since the above was written, and at the time I am preparing the second edition of this Essay) been for some time employed on experiments with a view to settle the above question. Whether I shall be able to devise a simple apparatus to effect the manufacture of Tea without charcoal is doubtful, but I can, I think, now safely affirm that the fumes of charcoal are *not* necessary to make Tea. On this point I am myself quite satisfied. The advantages of making Tea with any fuel (wood, coal, or anything else) would be numerous.

1.—Economy.
2.—Absence of charcoal fumes.
3.—Less chance of fire in Tea Houses.
4.—Probably reduced temperature in Factories.
5.—Great saving of labor.
6.—Saving of fuel—for it takes much wood to make a given weight of charcoal.

120 MANUFACTURE.

In addition to all the above, the wholesale destruction of forests that now takes place in all Tea Districts, in order to supply the charcoal for Tea, would be much lessened.*

I have seen a machine advertised for packing Tea, that is to say for so pressing it down that a large quantity shall go into a chest. I have never seen the machine and so cannot say how it works, but I do not think such a machine at all necessary. By the mode of packing, described at page 149, as much Tea as a chest will hold, *with safety*, can be put into it. If more were forced in, the chest would probably come to pieces in transit.

I see a sifting machine is now being advertised. "Jackson's sifting machine." I have seen drawings of it, but not the machine itself. In the one respect, that it is much larger than any thing used hitherto, it is more likely to succeed.

There is a machine for sifting and fanning Tea at one and the same time. I know not who invented it. It is a simple winnowing machine with sieves placed in front of the fan. By means of a rod and crank attached to the axle of the revolving fan the sieves are made to shake from side to side when the fanners are turned. The Tea is put into the upper sieve, a coarse one, and passing successively through finer ones, is thus sorted into different Teas. The open leaf at the same time is blown out by the fan.

I purchased one, but I do not find it does the work well. Sifting Tea is a nice process, and I did not find it sorted the Teas with any nicety. I have taken out the sieves, and use it now only for fanning, which it does very well, though no better than an apparatus which could be constructed at one-third the cost.

I do not believe in *any* present or future machine for sifting Tea, inasmuch as it is an operation, which to be well

---

* See this subject further discussed in the Addenda.

done, has to be continually varied. More will be said on this head further on.

I have now detailed shortly all the Tea machines or contrivances I know, or have heard of, and in my opinion Tea manufacture has not been much benefitted by any. I must however except rolling machines for these, as stated, are very useful to partially roll-leaf, but in my opinion it must be *finished* by the hand.* There is plenty of room yet for inventors. The machine, as before observed most to be desired, is one to separate the small Pekoe leaves from the others, ere the rolling of the leaf is commenced. If such a machine existed, it would much increase the value of all Indian Teas, and if the Agricultural and Horticultural Society are inclined to offer a prize for any machine it should be this.

At the point, where the separation should take place, the stalk is much tenderer than elsewhere, and this led me to think a blow or concussion, on the mass of green leaf might effect the object. I attached a bow, by the centre to an immoveable board, placed at right angles to the plane of a table (like the back of a dressing table) and then, causing leaf to drop from labove, subjected it to sharp strokes from the string of the bow. It effected the object partially, for many Pekoe ends were detached, but it bruised and cut the other leaf too much also. I believe a revolving barrel, with blunt, but thin narrow iron plates inside, which would strike the leaf placed within, as the barrel was turned, would perhaps answer. I give the above idea for what it is worth, for any inventive genius to improve on.

As it is impossible, as far as I can see, to construct any machine, which should *cut* the stalk *only* in the right place, *ergo*, I believe some arrangement which would take advantage of the fact, that the stalk is tenderer there than elsewhere, is the only one that could answer.

---

* I now believe Jackson's rolling machine, previously alluded to, will finish the rolling entirely.

Now to return to the manufacture of Tea. I will consider each of the five operations detailed, which I believe are all that are necessary to make good Tea, separately.

Withering.—There are several tests to show when leaf is withered. Fresh leaf squeezed in the hand, held near the ear, crackles, but no sound should be heard from withered leaf. Again, fresh leaf, pressed together in the palm of the hand, when released, springs back to nearly its original bulk, but withered leaf, in like circumstances, retains the shape into which it has been pressed. The stalk of withered leaf will bend double, without breaking, but fresh leaf stalks, if bent very little, break. Practice though soon gives a test superior to all these, viz. the feel of the leaf. Properly withered leaves are like old rags to lay hold of, and no further test after a time, than the feel of the leaf is necessary.

The agents for withering leaf are sun, light, heat, and air. Of these the most powerful is sun, for it combines all the others with it. Light is a powerful agent, for if some leaf be placed in a partially dark room, and some in a well-lighted verandah, the latter will wither in half the time the former will take. If light and moderate ventilation be present, heat is a great accessory to rapid withering.

There is often great difficulty in withering leaf in the rains. It *can* be withered in Tea pans, but "the out-turn" is then more or less injured, for after infusion the out-turn comes out green instead of the proper "new penny" color. Withering in dholes is also objectionable for the same reason, though if the heat is moderate the green effect is less. It is further a long and tedious operation.

Space and light are the great wants for withering leaf in wet weather. Bamboo mechans, tier above tier, should be constructed in every available space. Large frames, covered with wire mesh, may also be made (by means of

MANUFACTURE. 123

weights running over pulleys) to run up to the roof of any Tea building. The leaf withers well in such frames, for heat ascends, and much heat is given out by dholes.

It signifies not though where leaf is spread as long as there is space and light. Houses made of iron and glass would be far the best for withering leaf, for, if well ventilated, all the necessary agents for withering, detailed in the last page, would be present. I do not doubt the day will come when these will be used, for properly withered leaf is a necessity for good Tea.

In dry weather, when leaf comes in from the garden, spread it thinly anywhere and turn it once early in the night. It will generally be withered and ready to roll next morning. If not quite ready then put it outside in the sun. Half an hour's sunning will probably finish it.

In wet weather, if there is any sun when it comes in, or any time that day, take advantage of the sun to wither the leaf *partly*, so much that, with the after withering all night under cover, it will be ready next morning. If not ready next morning put it out in the sun, if there is any, till it *is* ready.

In very wet and cloudy weather, when there is no sun and continual rain, so that the leaf *cannot* be put outside (for remember that outside, when there is no sun, the light alone will wither it) artificial withering of some kind must be resorted to. I have mentioned the only means I know of for doing this.

As properly withered leaf is an important point in making good Tea, it is well worth while to keep one or two men, according to the quantity of leaf, for that work alone. They soon learn the best way to do it, and if made answerable the leaf is properly ready for the rollers, the object is generally attained. In this and every thing else in Tea manufacture, give different men different departments, and make them

R

answerable. Much trouble to the manager, who should supervise all, and much loss to the proprietor from bad Tea, will then be avoided.

Rolling.—This is a simple operation enough when the men have got the knack of it. Some planters advocate a circular motion of the hands when rolling, under the impression it gives the leaf a better twist. Some like rolling it forward, but bringing it back without letting it turn during the backward motion. I believe in neither way, for it appears to me to be rolled no better, or no worse, by these plans than by the ordinary and quicker mode of simply rolling it *any way*. The forward and backward motion is the simplest and quickest, and the way all rollers adopt, who are given a certain quantity of leaf (say 30 lbs. a fair amount) to roll for their day's work. In this ordinary rolling the ball in the hands, 'tis true, does not turn much in the backward motion, for 'tis more or less *pulled* back, but whether it turns or not, does not, I believe, signify the least.

Rolling in hot pans was formerly extensively practised. It is not much done now. I have tried the plan, but found no advantage in it.

Rolling on coarse mats, placed on the floor, might be seen also. When I visited the Assam Company's gardens near Nazerah in Assam, I saw it done there. It is a great mistake. The coarse bamboo mat breaks the leaf sadly, and much of the sap or juice from the leaf, which adds much to the strength of the Tea, runs through the coarse mat, and is lost.

One and the principal reason why Indian Tea is stronger than China is that in India the juice or sap is generally retained, while in China it is strange to say purposely wasted.

A strong immovable smooth table, with the planks of which it is formed well joined together, so that no apertures exist for the juice of the leaf to run through, is the best thing to roll on. If covered with a fine sectul pattic mat, nailed down over the edges of the table, a still greater security is given against the loss of any sap, and I believe the slightly rough surface of the mat enables the leaf to roll better. An edging of wood, one inch above the surface of the table should be screwed on to the edges over the mat, if there is one to prevent leaf falling off.

The leaf is rolled by a line of men on each side of such a table (four-and-a-half feet is a good width for it) passing up from man to man, from the bottom of the table to the top. The passage of each handful of roll, from man to man, is regulated by the man at the end, who, when the roll in his hand is ready, that is, rolled enough, forms it into a tight compressed ball (a truncated shape is the most convenient) and puts it away on any adjacent stand. When he does this, the roll each man has passes up one step.

The roll is ready to make up into a ball, when it is in a soft plashy state, and when in the act of rolling it gives out juice freely. None of this juice must be lost, it must be mopped up into the roll, again and again in its passage up the table, and finally into the ball, when made up.

There will be some coarse leaves in the roll which cannot be twisted. These, if left, would give much red leaf in the Tea. They should be picked out by, say, the third or fourth man from the head of the table, for it is only when the leaf has been partly rolled that they show. The man who picks out the coarse leaf should not roll at all. He should spread the roll, and pick out as much as he can, between the time of receiving and passing it on. In no case allow roll to accumulate by him, for if so kept it hardens and dries, and gives extra work to the last rollers to bring it into the mashy state again. Besides which, I rather think, any such lengthened stoppage in the rolling helps to destroy Pekoe ends, and is certainly injurious to the perfect after-fermentation, inasmuch as it, the fermentation partly takes place then.

This finishes the rolling process. Each man as stated can do 30 lbs., but there is further work for him to be now described.

Fermenting.—The balls accumulated are allowed to stand until fermented. I look on this being done to the right extent and no more, as perhaps the most important point in the whole manufacture.

Some planters collect the roll after rolling in a basket, and there let it ferment, instead of making it up into balls for that purpose as described. I much prefer the ball system for the following reasons:—When a quantity is put into a basket together and allowed to ferment a certain time, what was put in first is naturally more fermented than what was put in last, the former probably over, the latter under-done. The balls on the contrary can be each taken in succession *in the order they were laid on the table,* and thus each receive the same amount of fermentation. I think further the twist in the leaf is better perserved by the ball plan, and also that a large quantity in a basket is apt to ferment too much in the centre.

It is impossible to describe, so that practical use shall be made of it *when* the balls are sufficiently fermented. The outside of the ball is no good criterion. It varies much in color, effected by the extent the leaf was withered.* You must judge by the inside.

Perhaps as good a rule as any is that half the twisted leaves inside shall be a rusty red, half of them green. Practice alone,

---

* The more the leaf is withered the thicker in consistency and the smaller in quantity the juice that exudes, as also the yellower in color. Further, the more the leaf is withered the darker the outside of the balls. Bright rusty red is the color produced with moderately withered leaf. Very dark greenish red with much withered leaf.

however, will enable you to pronounce when the balls are properly fermented. There is no time to be fixed for it. The process is quicker in warm than cool weather.

The fermentation should be stopped in *each* ball just at the right time. Great exactitude in this is all important, and therefore, as I say, the balls should be taken in rotation as they were laid down.

The fermentation is stopped by breaking up the ball. The roll is spread out *very* thin, and at the same time any remaining coarse leaves are picked out.

This concludes the fermenting process.

Sunning.—The roll is then without *any* delay put out in the sun, spread *very* thin on dhallas or mats. When it has become blackish in color it is collected and re-spread, so that the whole of it shall be effected by the sun. With bright sunshine an hour or even less suns it sufficiently. It is then at once placed in the dholes, which must be all ready to receive it.

If the weather is wet, it must *directly*, the balls are broken up, and the coarse leaf is picked out, be sent to the dholes. This is the only plan in wet weather, but the best Tea is made in fine weather.

Firing or Dholing.—In the case of wet weather, unless you have very many dholes, fresh roll will come in long before the first is finished. The only plan in this case is to half do it. Half-fired the roll does not injure with *any* delay, but even half an hour's delay, between breaking up the balls and commencing to drive off the moisture, is hurtful.

In any but wet weather necessitating it the roll can be fired at one time, that is, not removed from the drawer, until it has become Tea.

The roll in each drawer must be shaken up and re-spread, two or three times, in the process of firing. The drawer must be taken off the fire to do this, or some of the roll would fall through into the fire, and the smoke thus engendered would be hurtful. If the lowest drawer is made to slide in and out a frame work covered with zinc should be made to run into a groove below it, and this zinc protector should be always run in before the lower drawer is moved. This is part of Mr. McMeekin's invention, and is very necessary to prevent roll from the lowest drawer falling into the fire when it, the lower drawer, is moved.

The roll remains in the drawers, subject to the heat of the charcoal below, until it is quite dry and crisp. Any piece then taken between the fingers should break with the slightest attempt to bend it.

The manufacture is now completed. The roll has become Tea.

All the above operations should be carefully conducted, but I believe the secret of good Tea consists simply in, *first*, stopping the fermentation at the right moment; and, *secondly*, in commencing to drive off the moisture immediately after.

I do not say that the manufacture here detailed may not be improved upon later, but I do say that in the results of economy, strong liquor, and well-twisted leaf, its results are very satisfactory, and not surpassed by any other mode at present in vogue. I do not pretend that it will give Teas rich in Pekoe tips. To attain this light rolling as shown must be resorted to, but just as far as Pekoe tips are procured so far must strength be sacrificed. Until the small Pekoe leaves can be detached and manufactured separately, this must always be the case.

From the Tea made, as described by sifting and sorting, all the ordinary black Teas of commerce, as detailed at page 136, can be produced, excepting "Flowery Pekoe."

To make Flowery Pekoe the closed bud and the one open leaf of the shoot are alone taken, and these are manufactured alone. It does not, as a rule, pay to make this Tea at all, though it fetches a long price. It does not pay for the following reasons:—

1. After the head of the flush is taken the pickers that follow do not readily recognize the remainder of the shoot, and consequently omit to pick many of them. A heavy loss in the yield is thus entailed.

2. The after Teas, made without these small leaves, are very inferior, as they are much weaker, and totally devoid of Pekoe tips.

3. The labor, and *ergo* the expense of picking the flush, is double.

The manufacture of Flowery Pekoe is simple enough. When the two leaves from each shoot of which it is made are collected they are exposed to the sun, spread out very thin, until they have well shrivelled. They are then placed over small and slow charcoal fires, and so roasted very slowly. If the above is well done, the Pekoe tips (and there is little else) come out a whitish orange color. The whiter they are the better. If the leaf is rolled *very lightly* by the hand *before* sunning, the liquor will be darker and stronger, but the color of the tips will not be so good.

Flowery Pekoe is quite a fancy Tea, and for the reasons given above it can never pay to make it.

Green Tea.

The pans for this should be 2' 9" diameter and 11' in depth. They should be thick pans, which will not, therefore, cool quickly. Many are required for this manufacture, 4 or 5 for every maund of Tea to be made daily. They should be set up in a sloping position, and the arrangement of the fire places such that the wood to burn under them can be put in through apertures leading into the verandah. One chimney will do for every two pans, and it should be built high so as to give a good draft, for hot fires are necessary.

Flat bladed sticks are used to stir both the leaf and the Tea in the pans, for the hand cannot bear the heat.

The men when working the Tea in the pans should have high stools to sit on, for it is a nine hours' job.

The bags in which "the roll" is placed at night should be made of No. 3 canvas, 2 feet long and 1 foot broad.

I will now detail the manufacture.

To make Green Tea the leaf must be brought in twice in the day. What comes in at one o'clock is partly made the same day. The evening leaf is left till the following morning, laying it thick (say 6 inches), so that it will *not* wither. But if the one o'clock, or the evening leaf comes in wet, they must both be dried, the former *before* being put into the pans the latter *before* being laid out for the night.

The manufacture thus begins twice daily, *viz.*, morning and one o'clock, but "the roll" of both these are treated together up to the time "the roll" is ready to place in the bags.

The leaf having no moisture in it is placed first in hot pans, at a temperature of say 160° and stirred, with sticks for about seven minutes, until it becomes moist and sticky. It is then too hot to hold long in the hand.

It is then rolled for two or three minutes only on a table until it gets a little twisted.

Then lay it out on dhallas in the sun (say 2 inches thick) for about three hours, and roll it thrice during that time, always in the sun. It is ready to roll each time when "the roll" has become blackish on the surface. It is not rolled more than three minutes each time, and then spread out as before. If you put on a proper number of men to do this they do each dhalla in succession, and when they have done the last,

MANUFACTURE. 131

"the roll" in the first dhallas will be blackish on the surface again, and ready to roll again.

When three rollings are done the roll should have a good twist on it.

It is then placed in the pans, at the same heat as before, and worked with sticks as before for two or three minutes, until it becomes too hot to hold.

It is then stuffed, as tight as it *can* be stuffed, into the bags described above, putting as much into each bag as you can possibly get it to hold. The mouth is then tied up and the bag beaten with a flat heavy stick to consolidate the mass inside, and so it is left for the night.

Next morning it is taken out of the bags, and worked with the flat sticks as before in the pans for nine hours without intermission. The temperature 160° at first down to 120° at the last.

During this last process the green color is produced, and the Tea is made. It is worked quicker and quicker as the hours pass.

The following are the kinds of Tea into which it is best sorted :—

1. Ends
2. Young Hyson
3. Hyson
4. Gunpowder
5. Dust
6. Imperial

Their relative value is in the order in which they are numbered.

The sorting of Green Tea is a nicer operation, and take, twice as long as sorting Black Tea.

If there are pans enough, and the work is well arranged, there should be no night work with Green Tea, for all should be over by 5 P.M. Whereas with Black Tea night work is generally a necessity.

The price obtained for Green Tea is more dependant on its appearance than in the case of black.

s

It is not easy to make Black and Green Tea in the same factory.

Green Tea, if well made, pays much better than Black Tea; and as before observed I think all gardens with China plants should adopt the manufacture. When once the building is fitted for it, and the routine established, the Green Tea manufacture is always preferred by those who have tried both.

The Hybrid plant makes the best Black, the China the best Green Tea. Would it not then be well that each planter made what he can make best? In Hazareebagh, in Kumaon, in the North-West generally, they are not blind to the above, how is it that Darjeeling alone refuses to change?

CHAPTER XXV.

SIFTING AND SORTING.

This is a very important item in the manufacture of Tea. Careful and judicious sifting, as contrasted with the reverse, may make a difference of two or three annas a ℔. in the sale of Teas.

I was shown some Tea, quite lately, which, as regards "liquor," was valued by the brokers at Re. 1-3 per ℔., but the "Tea" at only 14 annas! This was entirely owing to faulty sifting and sorting.

I don't believe in *any* machine for Tea-sifting, simply because it is not a regular process.* For example, you cannot say that, to make Pekoe, you must first use one sieve, then another, and so on. The sizes of sieves to be used, and the order in which they are to be used, will vary continually, as both are decided by varying causes, viz., the comparative fineness or coarseness of the Tea made daily, the greater or less presence of red leaf in it, and (because Tea varies much during the season and gets coarse towards the end) by the time of the year. These points all necessitate changes in the sizes, and the order of the sieves.

'Tis true sieves might be changed in a machine as required, but the only machine that could even pretend to save labor, would be one in which all the sieves were arranged one below the other, and thus the Tea would fall through each alternately, the motion being common to all. But this won't do for Tea sifting. Judgment must be used to decide *the length of time* each sieve is to be shaken; further, with *how much motion* it shall be shaken, &c., &c. But this is simply impossible with any machine, though all necessary to sift Tea well.

---

* We have yet to see what Jackson's Machine advertized can do.

The cost of Tea-sifting by hand (see page 161) is not eight annas *per maund*, including picking out red leaf, which *must* be hand work. Good and bad sifting will affect the value three annas per ℔. or Rs. 15 per maund!

With all parts of Tea manufacture it is well to employ the same men continually in each department, but above all, perhaps, should this be done in Tea-sifting. A good sifter is a valuable man. He knows each kind of Tea by name; he knows what sieves to use, and the order in which to use them for each Tea; what the effect a larger, or smaller mesh, will have on each kind, &c., &c., &c. In fact, he knows much more of the *practical* part of sifting than his master can, though the latter is, probably, a better judge how far the Teas are perfect when made.

Tea sieves are of two kinds, both round. One made of brass wire with wooden sides, 3½ inches high, the other cane, with bamboo sides, 1¼ inches high only. The latter are called "Chinese sieves," and though the brass ones are used in many places, there is no possible comparison between them, for the labor required in the use of the brass ones is much greater, and the results, as regards well-sorted Tea, much better with the Chinese.

Both kinds are numbered according to the number of orifices in the mesh contained in one linear inch. Thus a No. 6 sieve has six orifices to the inch in both; but in the brass kind, a No. 6 has six orifices *including* the wire; in the China kind, the cane between each aperture is *not* included in the measure. Thus the orifice in a No. 6 China sieve is exactly ⅙ of an inch square, but somewhat less in a brass sieve.

As I well know brass sieves cannot remain in favor after the others have been only once tried, I shall confine my directions to the China kind.

I practise, and I advise, Tea to be sifted daily. The Tea

SORTING. 135

made one day, sifted the day after, and in fact stored away in the bins ready sifted. I find it is more carefully done this way, for by the other plan a larger quantity being done at once by several men, they cannot from want of practice be expert. But by the daily plan, one, two, or three men as necessary, can always be kept on the work, and consequently they learn, and do it well.

To sift the following China sieves are required, and if daily sifting is resorted to, they will be found ample for any ordinary-sized garden:—

* 4 of No. 4
6 of No. 6
6 of No. 7

9 of No. 9
9 of No. 10
6 of No. 12

4 of No. 16.

Previous to sifting all red leaf should be picked out of the Tea. This, as stated under the head of Manufacture, should be done twice before the "roll" is fired, but towards the end of the season especially, some will still remain in the made Tea, and this must be carefully separated.

From what I have said it is evident that no rules can be laid down as to what sieves to employ to get out certain Teas. Only practice can teach this.

Further practice can only enable you to judge in a Tea Broker's point of view of different classes of Tea. This essay would, however, be incomplete, did it not contain a description of these. Such a description has been ably given by Mr. J. H. Haworth in his "Information and advice for the Tea-planter from the English market," (*Journal, A. & H. Society of India, vol. XIV*), and, as his knowledge on the subject is far in advance of mine, and consequently more to the point than any description I could give, I will close

---

* Even to break Tea on them it is a mistake to use brass sieves. Tea is best broken by a wooden roller, heavily weighted with lead, run in. The glaze or gloss on the Tea is thus preserved.

this chapter with the following extract from his valuable pamphlet, and trust he will excuse my doing so:—

Of the different classes of Tea.

Teas are arranged in various classes according to the size, make, and color of the leaf. I treat first and principally of the Black descriptions, as Green Teas are manufactured in only a few of the Tea-growing districts of India.

The following classes come under the name of Black Tea:—

Flowery Pekoe.
Orange Pekoe.
Pekoe.
Pekoe Souchong.
Souchong.
Congou.
Bohea.

The various broken kinds, viz.:
- Broken Pekoe.
- Pekoe Dust.
- Broken mixed Tea.
- Broken Souchong.
- Broken Leaf.
- Fannings.
- Dust.

We occasionally meet with other names, but they are generally original, and ought not to be encouraged, as a few simple terms like the above are sufficiently comprehensive to describe all classes manufactured.

Perhaps before entering into a detailed description of the various classes it will be well to explain the term "Pekoe" (pronounced Pek-oh), which as we see occurs in so many of the names above quoted. It is said to be derived from the Chinese words "Pak Ho" which are said to signify white down. The raw material constituting Pekoe when manufactured is the young bud just shooting forth, or the young leaf just expanded, which on minute examination will be found to be covered with a whitish velvety down. On firing these young leaves the down simply undergoes a slight change in color to grey or greyish yellow, sometimes as far as a yellowish orange tint.

When the prepared Tea consists entirely of greyish or greenish greyish Pekoe, with no or very little dark leaf mixed, it is called Flowery Pekoe.

Flowery Pekoe is picked from the shrub entirely separate from the other descriptions of Tea, only the buds and young leaves being taken. In the preparation it is not subjected so severely to the action of heat as the other classes of Tea, and generally preserves a uniform greenish grey or silvery grey tint. Its strength in liquor is very great, flavor more approaching that of Green Teas, but infinitely superior, having the strength and astringency without the bitterness of the green descriptions. The liquor is pale, similar to that of Green Tea, and the infused leaf is of a uniform green hue. In many instances where too much heat has been employed we find dark leaves intermixed, and the prevailing color, green, is sprinkled with leaves of a salmony brown tinge which is the proper color for the out-turn of any other ordinary black leaf Tea. A very common mistake is to call an ordinary Pekoe that may contain an extra amount of Pekoe ends, Flowery Pekoe. When this class of Tea is strong and of Flowery Pekoe flavor, it is called by the trade a Pekoe of Flowery Pekoe kind. In England Flowery Pekoe sells, as a rule, from 4s. 6d. to 6s. 6d. per ℔. One parcel has sold as high as 7s. 6d.

By many people the expediency of making Flowery Pekoe is much doubted. The true Flowery Pekoe leaf is the one undeveloped bud at the end of each twig. To pick this alone, without any ordinary Pekoe leaves, involves a great deal of trouble and expense, and I think, though the Flowery Pekoe be very valuable, that the account would hardly balance when we consider the deterioration of the Pekoe by the abstraction of the young leaves.

An ordinary Pekoe is a Tea of blackish or greyish blackish aspect, but dotted over with greyish or yellowish leaves which, on close inspection, will be found to possess the downy appearance which gives the name to Pekoe. In general we do not find the whole leaf covered with down, but only part of it, which in its growth has been developed later than the other parts. These are called by the trade "Pekoe ends," when very small Pekoe tips. A Pekoe is generally of good to fine flavor, and very strong, and its liquor dark. Its value is from 2s. 9d. to 3s. 8d. per ℔.

When the Pekoe ends are of yellowish or orange hue, and the leaf is very small and even, the Tea is called Orange Pekoe. In flavor it is much the same as an ordinary Pekoe, and many growers do not separate the two varieties, but send them away in the finished state mixed together. Its value is from 2*d*. to 4*d*. per ℔. more than Pekoe.

The term Pekoe Souchong is generally applied to a Pekoe that is deficient in Pekoe ends or to a bold, Souchong class leaf, with a few ends mixed. We often meet with it applied to an unassorted Tea, including perhaps Souchong, Congou, a few Pekoe ends, and some broken leafs. Prices range from 2*s*. 3*d*. to 2*s*. 10*d*.

The name of Broken Pekoe indicates at once what class of Tea it is, namely, Pekoe which has been broken in the manipulation or otherwise. It possesses the strength and fine flavor of a full leaf Pekoe, being therefore only inferior to it in point of leaf. In value it is very little inferior to Pekoe, sometimes as valuable, or even more so, as owing to the frangibility of the tender Pekoe ends, they are sometimes broken off in very large quantity, thus adding to the value of the broken Tea, though at the same time deteriorating the Pekoe. Prices from 2*s*. 6*d*. to 3*s*. 4*d*.

Pekoe dust is again still smaller broken, so small in fact as actually to resemble dust. It is of great strength, though often not pure in flavor, as frequently any dust or sweepings from other Tea is mixed with it to make the lot larger. The price of Pekoe dust may range from 1*s*. 6*d*. to 2*s*. 8*d*.

A tea only slightly broken is often called by the planter Pekoe dust; again an Orange Pekoe is often called Broken Pekoe, and the converse. A knowledge of the signification of these and other terms would teach the grower to be very careful in marking his Teas, as the nomenclature influences to a great extent the sale in the home market.

Having described the finer Teas we now come to the consideration of the classes of Tea which form the bulk of the manufacture of a garden.

Souchong may be taken as the medium quality, and when experience and skilled labor are employed in the manufacture as the bulk of the produce of an estate. The qualifications for being comprehended under this term are just simply an even, straight, or slightly curled leaf, in length varying say from half an inch to one-and-a-half inch. It has not the deep strength of Pekoe, but is generally of good flavor and of fair strength. The prices of Souchong are from 1s. 10d. to 2s. 8d.

Congou comes next. It may be either a leaf of Souchong kind, but too large to come under that class, or though of smallish sized leaf, too unevenly made, or too much curled (so as to resemble little balls) to be so classified. The flavor is much the same as that of Souchong, but the Tea has not so much strength. Some of the lower and large leaf kinds may be only worth perhaps from 1s. 3d., to 1s. 6d., whereas the finer qualities sell as high as 2s. to 2s. 3d. per ℔.

Bohea is again lower than a Congou. It may be either of too large a leaf to be called Congou, or, as is generally the case, it may consist principally of old leaf, which on being fired does not attain the greyish blackish color which is so desirable for all the black leaf kinds except Flowery Pekoe, but remain of a brownish or even pale yellowish hue. It has scarcely any strength, and is generally of coarse flavor, sometimes not, but is never of much value unless of *namuna* kind (a term which will be described hereafter.) We may quote prices at from 3d. to 1s. 2d. per ℔.

We now come to the broken descriptions of these middle and lower classes of Tea.

Broken mixed Tea is, as its name imports, a mixture of the various kinds of Tea broken. It may have a very wide range, include some of the lower classes or approach Broken Pekoe in character and value, but the kind usually thus named is a Tea worth from 1s. 8d. to 2s. 6d. generally of a blackish aspect, and containing a few Pekoe ends.

The term Broken Souchong is commonly and appropriately applied to a Tea, which, though broken, has some approach

T

to a full leaf, and that of the even Souchong character. Its value may vary, say from 1s. 6d. to 2s. 2d.

Broken leaf is a term of great comprehensiveness, but generally is used to signify a Tea worth from 8d. to 1s. 1d. per ℔. It may be of a brownish, brownish blackish or blackish color. Its strength is seldom great, but its flavor may be fair or good, but in the lower qualities it is generally poor, thin, or coarse. It would be better to employ this term only as a general name of Broken Tea, and not to use it to signify any particular class, as it is very indefinite.

Fannings is similar in color and class of leaf to broken leaf as described above; in value also much the same, perhaps on the average a little lower. I suppose, in most cases, the mode of its separation from the other classes of Tea is, as its name implies, by fanning.

Dust is a very small broken Tea, so small in fact as to approach the minuteness of actual dust. It is often very coarse, or "earthy" in flavor, owing perhaps to sweepings and dust having become mixed with it. Its value is from 6d. to 1s. 6d. In any Tea of this class worth more than these quotations, a few Pekoe ends or tips will generally be found, which bring it under the name of Pekoe Dust.

We will now look at Black Teas in a body, and point out what is desirable and what is objectionable in them.

We have seen that all Teas which contain Pekoe fetch higher prices than others, consequently we infer that Pekoe is a desideratum. If we glance at the descriptions of the various classes of Tea which have been given above, we shall find that it is an element of strength and good flavor. I do not mean to say that any Pekoe is stronger or of better flavor than any Tea which does not contain Pekoe, as the soil, the climate, the cultivation, the manufacture and various other causes, may influence the strength and flavor of different Teas; but, as a rule, in Teas that are produced under the same circumstances, the classes containing Pekoe are stronger and of better flavor than those without it.

There is another class of Tea which I have not yet

described that possesses very great strength and very fine flavor. This is the class known as the "namuna" kind. All readers of these pages who have been connected with India any time will recognise the word* though they may not quite see how it comes to occupy the position in which we consider it. It is said that its first application in this manner arose from a planter having sent to England some sample boxes of Tea with the ticket "namuna" on them. These Teas happened to be of the peculiar description which now goes by that name, and which I proceed to describe. The London Brokers have always since then applied the name namuna to this class of Tea. The leaf may have perhaps the ordinary greyish blackish aspect, with generally a greenish tinge. In the pot it produces a very pale liquor, but on tasting it its quality belies the poor thin appearance of the infusion. It is very strong, stronger by far than ordinary Pekoe; in flavor say about half way between a Flowery Pekoe and a Green Tea, quite distinct from the Flowery Pekoe flavor, possessing somewhat of the rasping bitterness of the Green Tea class with the flavor a little refined. The out-turn is generally green, sometimes has some brownish leaves mixed. Any of the black leaf Teas may be of this class from the Pekoe to the lowest dust, and all throughout the scale, if the flavor be distinct and pure, may have their value enhanced from 4*d*. to 10*d*. per ℔.

Similar in every respect, except one, is the Oolong kind. The one wanting quality is the strength, sometimes, by-the-bye, the flavor is a little different. It may have the greenish, greyish blackish leaf (though generally the green leaves are distinct from black ones, the Tea thus being composed of greyish, blackish leaves with a few green ones intermixed), always has the pale liquor, generally the greenish infused leaf, but sometimes it is sadly intermingled with black leaves, as it is a Tea whose flavor is frequently burnt out, though its

* I need hardly remark that the Hindustani word Namuna (pronounced Nemoonah) means sample.

weakness and green appearance are no doubt often caused by deficient firing. Teas of this kind on the average sell below the ordinarily-flavored Teas of the same class of leaf.

In Teas of ordinary flavor the following rules hold good:—The darker the liquor the stronger the Tea, and the nearer the approach of the color of the infused leaf to a uniform salmony brown, the purer the flavor. Whenever we see any black leaves mixed with it (the out-turn) the Tea has been overfired, and we may either expect to find the strength burnt out of it, or else to find it marred by having a burnt or smoky flavor incorporated with it. When you come across an altogether black or dirty dark brown out-turn, you may be certain of pale liquor containing little or no strength and no flavor to speak of, unless sometimes it be sour. This is a quality which I shall now touch upon, and regret that I cannot with any certainty give any reliable information whereby the planter may guard against this greatest of faults. It may have various grades,—slightly sourish, sourish and sour, depreciating the value of the Tea, say from 3d. to 1s. 6d. per ℔. The flavor of a sour Tea is hardly capable of description. It is not so acid as sour milk, in fact not acid at all, rather a sweet flavor than otherwise being blended with the sourness. It is extremely unpleasant in its more developed grades, and cannot be easily understood except by actual tasting. To the uninitiated this fault is only perceptible in the more strongly marked instances, but to one of the trade the least tendency to it not only condemns the parcel at once but also causes him to suspect any other lots made at the same or any other time by the same grower, as it is a curious but unaccountable fact that some two or three gardens (or growers?) almost always produce Teas having this fault. I will not cite all the different explanations that have been offered on this subject; I will simply quote the one which seems to have gained most ground, and leave those more competent than myself to express any opinion on the subject. The cause assigned to which I refer is that the Tea leaf after being picked is allowed

SORTING. 143

to remain too long in the raw state before being fired, during which time it undergoes a process of fermentation; some then say that this causes sourness, while others maintain that the fermentation is absolutely necessary for the production of a Black Tea. The fact that we never meet with sourness in a Green Tea, one feature in the preparation of which being that it is fired almost immediately on being gathered, goes to corroborate this view.

Burntness I have already referred to. As I said before it may either destroy the strength and flavor altogether, or sometimes without destroying the strength add an unpleasant burnt flavor to it. When the Tea has the flavor of smoke about it, it is called smoky or smoky burnt. By being burnt a Tea may be deteriorated in value, say from 2d. to 1s. per lb. The symptoms of burntness are a dead black leaf (as opposed to the greatly desired greyish, blackish color) having a burnt smell which often entirely neutralizes the natural aroma of the Tea. In looking over a broker's character of a parcel of Teas you may occasionally meet with the terms "fresh burnt," "brisk burnt," or "malty burnt." These phrases do not carry a condemnatory meaning with them. The meaning of the word burnt, as used here, would be better expressed by the term fired. The term malty means of full rich flavor, perhaps from the aroma of this class of Tea resembling somewhat that of malt. Teas of the three above descriptions, you may have noticed, often fetch very good prices. The meaning of the word "full" applied to a liquor is hardly appreciable except by tasting. It does not signify strength or flavor, but is opposed to thinness. A Green Tea may be strong or of good flavor, but its liquor is never full. Fulness is generally characterised by a dark liquor. The quality known as body in a wine is somewhat akin to fulness in a Tea. We speak of a "full" leaf Tea in contradistinction to a broken leaf. "Chaffy" is generally used in connection with Bohea and other brown leaf classes of Tea. A light (in weight) brown, open or flat leaf, in fact one resembling chaff, would be called chaffy. The lower classes of Tea, specially the dusts, are often described as "earthy" in flavor. By this a coarse low flavor is understood, perhaps often caused by the admixture of real dust.

When the make of a Tea is spoken of as a "well made," "fairly made," &c., leaf, the effect of the manipulation or rolling is referred to. We may have a "well made even," or a "well made mixed large and small," leaf. We may have a "straight" or "curled," or as the latter is generally expressed when applied to a large leaf Tea, "twisted" leaf. It may be "flattish-made," indicating that though the leaf is not open it wears a flattish aspect, or it may be open, which betrays a want of sufficient or skilful manipulation. A "wiry" leaf is small, perfectly rolled, and very thin (in diameter) generally rather curled, so as in fact to resemble small pieces of bent wire. It will be seen at once that only the finer Teas can have a wiry leaf, principally the Orange Pekoes and Pekoes. Some times we meet with a fine Souchong that may be thus described.

Green Teas.

As in the North-West Provinces Green Teas form the bulk of the produce, it will be well to give a short description of them, though the tenor of my remarks below will show the general opinion as to the desirability of making them.*

Gunpowder is the most valuable description, its price ranging from 2s. 8d. to 3s. 8d. per ℔. Instead of possessing the long and thin finished leaf which is the desideratum of Black Teas, it is rolled into little balls more or less round, varying from one-eighth to one-quarter of an inch in diameter. Sometimes it is not altogether composed of round leaf, but has some long leaf mixed.

---

* I think I need hardly pause to correct the popular error that the Green and Black Teas are made from two different species of plant. Most of my readers will know that they are both made from the same leaf, the difference lying only in the manufacture.

When the Tea is of the shape of gunpowder, but is larger than the size above quoted, it is called Imperial. Prices of Imperial are from 10*d.* to 2*s.* 6*d.*

Amongst Green Teas Hyson may be taken as the parallel of Souchong of the black leaf descriptions. Undoubtedly there is often much young Pekoe leaf in it, but all chance of *discriminating* it in the finished leaf is done away with by the change in color. Hysons sell from 1*s.* 2*d.* to 3*s.* 6*d.*

Young Hyson is smaller than Hyson, occasionally slightly broken. It fetches from 7*d.* to 2*s.* 6*d.*

Hyson skin consists of the bold broken leaf of Hyson and young Hyson. A small broken Green Tea is seldom sent on the home market. The reason of this is obvious. When we consider that Hyson skin only fetches from 7*d.* to 1*s.* it is apparent that anything approaching a dust would give very little chance of a profit. I have seen one or two parcels too much broken to come under the title of Hyson skin, sell at 3*d.* to 6*d.* per ℔. in London. It would be well if some of the Indian planters would take a lesson from the Chinese, and not send home their very low Teas, black or green, as they are very difficult of sale in London, and in many cases cannot pay the cost of packing and shipping. The Chinese make a great quantity of their broken Teas into Brick Tea and send it into the Central Provinces of Asia, where it meets with a ready sale. I do not see why this should not be done by the Indian growers. There is a large consumption of Tea on the other side of the Himalayas, not very far from Darjeeling and Assam. I hear also that in the neighbourhood of the growing districts, especially the North-West Provinces, the natives are beginning to consume largely, and will pay 8 as. to 1 Rupee for a Tea that could not possibly fetch more than 1*s.* to 1*s.* 6*d.* per ℔. in England. Whether the natives of India, as a whole, do or do not take to drinking Tea, will have a material effect on the future prospects of the article.

Before dropping the subject of Green Teas I will say a word or two as to the expediency of making Green Tea. I have questioned several experienced people on the subject, but none can tell me their especial object in manufacturing their leaf into Green Tea. One gentleman told me that he thought that it was because their Tea-makers (Chinamen) knew better how to make greens than blacks. I have carefully examined the leaf of several of the North-west Green Teas, and noticing their English sale prices consider that they would have sold on the average at least 3d. per ℔. higher had they been made into Black Tea. The best way to test this would be to have Green and a Black Tea made from the same leaf, and then to value the one against the other. I regret that I have never had the opportunity of doing this. We notice that the largest and most experienced producers never make Green Tea.*

I must not pass over Caper without a short description. It is a Tea which is made in large quantity in China, though I have only seen one parcel of Indian growth. It forms a link between the black and green descriptions. The color of the leaf is a very dark green, in form it is similar to a gunpowder, Imperial or round leaf Congou. The liquor is pale, and the outturn green, flavor perhaps nearer to that of a Green than of a Black Tea.

---

* This was written by Mr. Howorth some years ago. Later experience has, I think, proved he was mistaken. *Good* Green Teas sell to-day at far higher prices than Black, but the best can only be made from China plants.

CHAPTER XXVI.

BOXES, PACKING.

By far the best Tea boxes are the teak ones made at Rangoon. The wood is impervious to insects of all kinds, even white-ants. Sawn by machinery the pieces sent to compose each box are very regular. The plank is half inch, and each chest made up measures inside 23 by 18 by 18½ inches, and necessarily outside 24 by 19 by 19½ inches. The inner cubical contents are 7,659 cubic inches, and this suffices for above one maund of fine and under a maund of coarse Teas.

Each box is composed of fourteen pieces, *viz.*, for the two long sides three each, for the two short sides two each, two for the bottom, and two for the lid. By the arrangement of three pieces in the long sides, and two only in the short sides, the centre piece of each long side is attached to *both* the short end pieces, and thus great strength in the box is insured, there being no place where it can possibly separate at the joints.

These boxes are not made to "dovetail." Each piece (and they are sawn with mathematical regularity, as to length, breadth, and thickness) must be nailed to its neighbour. The best nails for this are the kind called "French Pins" 1¾ inches long.

The wood is sold at Rangoon in bundles, and could be landed in Calcutta for about Re. 1-8 or 1-12 per box. The boxes need not be made up till shortly before they are wanted, and in this form, of compact bundles of short pieces, are very convenient for transport and stowage.

Of course in many districts these boxes are not procurable, and local ones must be made. If so, use hard wood, and

148   PACKING.

make your boxes *about* the size given above, for small boxes add much to the cost of freight.

Let the planks be ⅜th of an inch thick, for ½ inch, that is ¼th inch boards are not strong enough, except they are of teak or any other very good wood.

Take care the joints of the several pieces, composing the sides and ends do *not* coincide at the corners, for if they do, the box is very apt to come asunder.

The best way to arrange the pieces is as described in the Rangoon boxes above.

"A form" must be made on which the inner leaden case shall be constructed. That is, a well-made, smooth box, to fit exactly into the box you pack in. It must be some three inches higher than the interior of the original box, and have bars running across inside, for handles to lift it up, and let the lead case slip off it, after it (the lead case) is finished.

Solder your lead case, over your form, in the way to waste least lead. In the Rangoon boxes described, two large, two small sheets* and one piece 22 by 9 inches (let in between the two large sheets) suffices, and there is little or no waste.

The lead case ready, hold up the form by the inner rods, and let the case slide off. Put it at once into the packing box, taking care no nails protrude inside, or anything else which will hurt it, and thus prepare all the boxes for the break of Tea you are about to pack.

One great advantage the Rangoon boxes, and in fact all machine-sawn boxes, have, is their equal or nearly equal weight. Purchasers of Teas, at the public auctions, require "the tare" of boxes to be as near the same weight as possible. If the tares differ much, say more than 3lbs., the Tea will be depreciated in value. It is well there should be *about* the same weight of Tea in all the boxes that contain

---

* Large lead is 37 by 22 inches. Small lead 25 by 19 inches.

any one kind, but this is not nearly so essential, as approximation in "the tares."*

Your boxes all ready and lined with lead, choose a fine day for packing. Do this whether you finally dry the Tea in the sun or over the dholes; for even in the latter case it is well to avoid a damp day.

But before you pack you must bulk. That is you must mix all the Tea, of any one kind, so intimately together that samples taken out of any number of chests shall agree exactly. This can be done by turning out all the Tea on a large cloth placed on the floor, and turning it over and over. No two days' Teas are exactly alike, and you have perhaps a month (say Pekoes) to pack. It is therefore necessary to mix them well.

Though I know many planters think the fumes of charcoal necessary and beneficial for the last drying, I do *not*. I have tried both sun and charcoal for it, and no difference was perceptible. The former costs nothing, is more commodious, and I always employ it when possible. The sun cannot burn the Teas, the charcoal, if the heat is too great, may.†

Whether you use sun or charcoal put the Tea hot into the boxes. The *only* object of the final drying is to drive off the moisture, which it will certainly, in a more or less degree, have imbibed since its manufacture. Even the large zinc-lined bins which should be fitted up in all Tea stores, and in which the Tea is placed after manufacture, will not entirely prevent damp, so in all cases a final drying is necessary.

Keep it in the sun, or over the charcoal, until it is hot throughout, hot enough to ensure all the moisture having been driven off. Then put into the box enough to about one-quarter fill it. Now let two men rock the box, over a half inch round iron bar, placed on the ground, until the Tea has

* I now put exactly the same weight into all the boxes holding any one kind of Tea.

† *If* I devise an apparatus for my furnace experiments, Tea will be best dried in that way when there is no sun.

well settled. Then place a piece of carpet over the Tea, the exact size of the box, and let a man stand inside and press it down a minute or two with his feet. Now fill up nearly another quarter, and press it again over the carpet as before. Repeat this, putting less and less into the box each time, as you near the top, until it is quite full, but do not rock it the last two or three times at all. Only press it with the feet as described. No patent screw press, or anything else, will pack the Tea better or more closely than this plan, and when the men are practised at it, you will find there will not be a difference of more than two or three lbs. in the Teas of any one kind put into the boxes.*

The box full, just even with the top, and well pressed down to the last, lay over the Tea a piece of the silver paper, which is found inserted between each sheet in the lead boxes. This prevents any solder or rosin getting on the Tea when soldering the top. Now fit on the lead sheet top, solder, and nail on the wooden lid.

Weight Tea in each box.—The boxes ready lined, with a lead cover loose, must be all weighed *before* the Tea is packed, and again *after* they are filled and soldered down, but *before* the wooden lid is put on. The difference of these weights, minus the weight of the little solder used in fastening down the top lead, (for which allow say one pound to give a margin also), will be the net weight of Tea in each box.

Thin iron hooping, put round both ends of the boxes, much increases their strength, and is not expensive.†

Stamp each box on its lid and on one end. Use for this zinc plates, with the necessary marks cut out in them. A brush run over these with the coloring matter does the work well and quickly.

---

* Best to weigh and put in exactly the same quantity. See note, p. 148.

† This should, except the lid part, be put on the boxes before the Tea is paced.

PACKING.

Let the stamp comprise the kind of Tea, the plantation or owner's mark, the number of the box, and the year, for instance—

Pekoe.
A
B   C
D
No. 80.   1871.

The invoice you send with the break must give for each box the number, the gross weight, the tare, the net Tea, and the kind of Tea, with a declaration at foot that the Teas of each kind have been, respectively, well bulked and mixed together before packing.

Remember the larger the quantity of Tea, of any one kind to be sold at one auction, the higher the price it will probably fetch. Sell, if possible, twenty or thirty chests of one kind of Tea at the same time, for small quantities, as a rule, sell below large both in Calcutta and London.

CHAPTER XXVII.

MANAGEMENT. ACCOUNTS. FORMS.

SYSTEM and order, a good memory, a good temper, firmness, attention to details, agricultural knowledge, industry, all these, combined with a thorough knowledge of Tea cultivation and Tea manufacture, are the requisites for the successful management of a Tea plantation.

To find men with *all* these qualities, is, I allow, not very easy, still they do exist, and such a one must be had if success in Tea is looked for.

Before the work is given out each day the manager should decide exactly what is most required, and apply it to that. He should write down, when distributing the men, the works and the number employed on each. This paper he should carry in his pocket, and he can then verify the men at work at each or any place when he visits it during the day.

The writer, the moonshee, and the jemadar, if there is one, should write similar papers when the coolies are mustered in the morning, and the manager should detail to each of these men which work they are particularly responsible for. This should also be shown in the "morning paper."

Each of the above men then measure out the work to the coolies. Visit it once or oftener in the day, and measure all that remains undone at night. A daily report of the work is kept, written by the writer in the evening.

The two forms, given below, are those I have adopted. The latter is suited to local labor paid daily, but it can easily be altered to suit either local labor paid monthly or imported coolies.

This is the Morning Paper.

Work to be done on 187 .

Detail.	No. of Garden.	In whose charge.	Headman on the Work.	Probable number.	Actual number.
			Total Coolies	...	

The column of "probable numbers" is given, so that before it is known exactly how many men will be present for work they can be divided in the most likely way.

Each head man (called "mate," "mangee," &c., in different districts) is best designated by a letter or number. In neither form would there be room to put in names at length.

The form below written in the evening is made into a book for each month. The advantages of it for after reference are great, and it has many other advantages too numerous to detail here, but which will be appreciated when it is used :

Work on   *for*   187 .

Detail of Work.	No. of Garden.	Mangees.	Chupp.		Coolies.	Measurement and Remarks. *Meting en aanmerk*
						As.
Total at work ...						× 3 =
⎯ Command ...						× 2½ =
Sick ...						× 2 =
Absent ...						× 1½ =
Total ...						× 1 =

Picking Leaf =

Making Tea =

Tea Sorting =

Cut =

Total =

The following is the plan I recommend for the leaf-picking and the Tea accounts.

The leaf of each picker is best measured in the field and

MANAGEMENT.

as loads are collected, brought to the factory by one or two men, throughout the day. It entails a loss of time, and further a depreciation in the leaf, if kept long in a close mass in one basket, each picker bringing his or her leaf to the factory twice a day. The pickers are paid so much per basket, holding in any case 2½lbs. I find the most convenient plan to give the mangee in charge of the pickers tickets of any kind for this, which tickets are changed for money in the evening. As each load of leaf comes in through the day it is weighed, and this gives a check on the tickets given by the mangee or mate. This is the meaning of the two columns in the form below "tickets by leaf" and "tickets paid."

In the form the first column of "leaf results" shows the condition of the leaf when picked whether wet (W) or dry (D). Unless this were noted the proper amount of Teas the leaf ought to make could not be known, and there would be no check against theft, which is carried on to a great extent in many gardens.

As explained previously, only the sections ready in each garden are picked. The sections are not entered in the form, only the number of the garden. The flushes now noted are the 20th, in some 21st, or 22nd in others.

The Tea is calculated from the leaf. It should be 25 per cent. if the leaf is picked dry and 22 per cent. if picked wet. As each load comes in a memorandum is made as to whether it is dry or wet, and the figures in the column "Tea should be" are thus found.

The Tea is weighed the morning after it is made and entered in the column "Tea made." The precentage it bears to the leaf is then calculated and entered in the account column.

After sifting the whole is weighed again, and the result entered in the column "Tea after sifting." Doing this is very important, for it checks theft. Directly after it is weighed this second time it is put in the bins in the store.

Daily Leaf and Tea Account.

Date.	TEA RESULTS.				LEAF RESULTS.							
	Tea should be.	Tea made.	Per cent.	Tea after it is sifted.	State of leaf.	Tickets.		Number of Gardens.	Flushes.			Total leaf.
						By leaf.	Paid.		20	21	22	
October Sunday, 1st	,,	,,	,,	,,	220 W. 600 D.	410	360	3 5 7 8	,, 310 112 ,,	170 ,, ,, ,,	,, ,, ,, 228	820
Monday, 2nd	198	200	24	199	D.	462	440	3 9	,, 410	515 ,,	,, ,,	925
Tuesday, 3rd	231	230	25	233	W.	200	180	1 2 3	430 ,, ,,	,, ,, 210	,, 160 ,,	800
Wednesday, 4th												
Thursday, 5th												
Friday, 6th												
Saturday, 7th												
Total for the week	,,	Mds. lbs. 16 33										

If this system is carried out no Tea (exceeding a ℔ or so) can be stolen, without its being at once missed, and the importance of this cannot be exaggerated. Tea proprietors do not guess *how* much is lost in this way. Maunds upon maunds might be stolen in many gardens, and unless the theft were accidentally discovered, there is nothing in the Tea accounts to show it to the manager.

I have suppositiously filled up the three first days of the form. The 820 lbs. leaf picked on Sunday is made into Tea on Monday. The 198 is written down Sunday evening. On Tuesday morning when the Tea is weighed and found to be 200 lbs., that is entered in the Monday line as also the percentage. On Tuesday evening after it is sifted and made into different Teas, it is weighed again and found to be 199 bs., and so entere d.

In dry weather after sifting, owing to dust flying off, it is always a little less. In wet weather, on the contrary, it increases in weight. In the Tuesday line where "W" shows lit was a wet day and the Tea 230 lbs. before sifting, is 233 afterwards. This is owing to moisture imbibed, and it is the only objection to sifting daily, whatever the weather. The advantages of the plan though are so great, as explained, that I put up with this, and practically I do not find it detrimental. Of course, as previously explained, all moisture is driven off before the Tea is *packed*. However, to make all quite safe, after a very wet damp day, the Teas might be redried for a few minutes over charcoal before being put into their respective bins. I do not do this myself though, and do not think it necessary.

I hope now I have made the above form plain. It is in a book and each page will hold one week. The total of the Tea made in the week is added up and shown at foot, and that amount is then transferred to the credit side of the Tea store account. Thus (see both forms) 16 maunds, 32 lbs. is credited.

The form given below is also kept in a book, and the total of the right hand side, subtracted from the left, gives at any time the quantity of Tea in store.

Tea Store Account.

Week ending on Saturday.		Tea made in week.		Total.		Date.	No. of Invoice.	To whom.	Tea in each Invoice.		Total.	
		Mds.	lbs.	Mds.	lbs.				Mds.	lbs.	Mds.	lbs.
Brought over	...	...	...	405	8			Brought over ...	...	...	351	14
October	7th	16	32			October	3	15	40	15		
„	14th	15	0			„	20	16	33	10		
„	21st	17	10								73	25
„	20th	14	40									
				63	2							
Carried over		...			...	Carried over			...			...

Regarding accounts between the manager and his employers I think they should be of the simplest kind. If a man *can* be trusted, he *should* be trusted, if he cannot, no system of accounts will restrain him, and he should be kicked out. A simple account current furnished monthly, showing under few heads the receipts and expenditure, is all that can be required. It is not by *any* papers received from a manager that an opinion can be expressed as to how he does his work, and how the plantation progresses. A competent person visiting the garden can easily ascertain, and in default of

MANAGEMENT.

this, and combined with this, the only true test is the Balance Sheet at the end of each season.

Shortly, it is not by the form, the nicety, the detail of accounts between manager and employer, that success is insured or even forwarded. It is, as far as accounts are concerned, by the forms and *system* the manager adopts as between him and his subordinates, and these he should be able to show are good to the employer, or any one deputed by him to visit the garden.

The profit shown yearly, whether it is large or small, all things considered, is however still the only true ultimate test.

CHAPTER XXVIII.

Cost of Manufacture, Packing, Transport, &c.

This is as follows:—It will vary more or less according to the site, rate of wages, &c., but in the form the tables are given, if not suitable to any case, it can easily be made so.

I have added Sorting, Packing, Freight to Calcutta, and Broker's Charges in Calcutta, to the cost, so that all is included from the moment the green leaf is picked off the trees till the hammer falls at the public auction.

Table cost of Manufacture, Sorting, Packing, Transport to Calcutta, and Broker's Charges for each maund of Tea.

	Rs.	As.	P.	Rs.	As.	P.
Manufacture.						
1 head man with the pickers, say				0	4	0
320 lbs. green leaf picked, at 1 pice per lb.*				5	0	0
1 man withering above leaf, at say 4 annas				0	4	0
¼ share head man in Rolling-house				0	2	0
10¾ men rolling above, at 30 lbs. leaf per man, and say 4 annas per man				2	10	8
¼ boy clearing out ashes of Dhole house, at say 2 annas				0	0	6
Carried over				8	5	2

---

* In practice the basket in which the leaf is measured being made to hold 2½ lbs., for which a ticket is given, representing 2 pice, the leaf to make a maund of Tea does not really cost so much.

COST OF MANUFACTURE.

	Rs.	As.	P.	Rs.	As.	P.
Brought forward ...	8	5	2			
¼ share head man in Dhole house	0	2	0			
1 man firing "Dhole work" say	0	4	0			
¾ maund charcoal for Dhole work, at 8 annas	0	6	0			
Lights for night work, viz., turning green leaf and dholing, say	0	4	0			
Wear and tear of dhallas, baskets, picking baskets, fuel for artificial withering, &c. ...	0	1	10			
				9	7	0
Sifting and Sorting.						
1½ boys to pick out red leaf, at say 2 annas	0	3	0			
1 sifting man, at say 4 annas ...	0	4	0			
Wear and tear of sieves, say ...	0	0	3			
				0	7	3
Packing.						
1 box	1	13	0			
4 sheets lead, viz., 2 large and 2 small	1	6	6			
Labor of lining box with lead, solder, closing lead, closing wooden box, stamping, and cost of nails	0	0	9			
Labor of drying previous to packing, whether in sun, or over Dholes, including charcoal, if the latter are used ...	0	0	9			
Carried over ...	3	5	0	9	14	3

COST OF MANUFACTURE.

	Rs. As. P.	Rs. As. P.
Brought forward ...	3 5 0	9 14 3
Labor of filling the box, shaking it well, and pressing down the Tea (2 men)	0 0 6	
		3 5 6

Transport.

Freight to Calcutta for one maund Tea, say	1 12 0	
		1 12 0

Broker's charges in Calcutta.

Landing, lotting, and advertising per chest	0 14 0	
Brokerage at 1 per cent. on the amount sale, say Rs. 70 for the maund ...	0 11 3	
		1 9 3
Total for one maund Tea ...		16 9 0*

N.B.—If more than two maunds Tea are made per day, some of the items under head of "Manufacture" would be a little less. See page 70 where it will be seen that each maund of Tea is worth to the manufacturer (after deducting all costs) Rs. 50.

---

* After experience has shown me this amount, when any quantity of Tea is made, is too high—Rs. 15 or 15·8 would be nearer the mark.

CHAPTER XXIX.

Cost of making a 300-acre Tea Garden.

In the following estimate 100 acres is supposed to be planted the first, 100 acres the second, and 100 acres the third year.

To elucidate a table I shall draw up in the next chapter showing the probable receipts and expenditure on such a garden for a series of years, I shall suppose this plantation to be begun in 1875, and number the years accordingly.

The expenditure would truly, in the supposed case, begin in the latter part of 1874, but it is more convenient to regard it as commencing 1st January 1875, and thus keep each year separate.

I estimate all new cultivation as planted "at stake," that is, the seed sown *in situ*. Nurseries are only to fill up vacancies.

I shall not pretend in this to go into minute details, such as are given at (page 83), for it is simply impossible to do so. The cost of making a plantation, must vary greatly, being determined by climate, available labor and its rates, lay of land, nature of jungle to clear, &c., &c. In this estimate only round numbers can be dealt with. The prices I assume are average ones, neither suited to very heavy jungle, and very expensive labor, or the reverse:

	Rs.	Rs.
1st year (1875).		
Purchase 700 acres land, at Rs. 8 per acre ...	5,600	
40 maunds seed, at Rs. 70 (¹)	2,800	
Nurseries for vacancies and labor transplanting (¹) ...	200	
First temporary buildings	1,000	
Carried over Rs. ...	9,600	

---

(1). The cost for seed, nurseries, and transplanting increases each year as the area, over which vacancies may exist, enlarges.

COST OF MAKING A GARDEN.

	Rs.	Rs.
Brought forward Rs. ...	9,600	
All expenditure to plant 100 acres, at Rs. 80 per Acre (²)	8,000	
Cultivating the said 100 acres first year, at Rs. 50 per acre (³)	5,000	
		22,600
2nd year (1876).		
60 maunds seed, at Rs. 70 (¹)	4,200	
Nurseries and labor transplanting (¹)	300	
Repairs, buildings and some new ones still of a temporary nature	500	
All expenditure to plant the second 100 acres, at Rs. 70 per acre (²)	7,000	
Cultivating first 100 acres, at Rs. 60, second 100 acres, at Rs. 50 per acre (³)	11,000	
		23,000
3rd year (1877).		
70 maunds seeds, at Rs. 70 (¹)	4,900	
Nurseries and labor transplanting (¹)	400	
Buildings for Tea manufacture (temporary) and repairs to buildings	3,000	
All expenditure to plant the third 100 acres, at Rs. 60 per acre (²)	6,000	
Cultivating first 100 acres, at Rs. 70, second at Rs. 60, third at Rs. 50 per acre (³)	18,000	
		32,300
Interest on first year's outlay two and a half years, second year's outlay one and a half years, third year's outlay half year, at Rs. 5 per cent. per annum	......	5,357
Total expense to make the 300-acre garden ...	......	83,257

The garden is now made at a cost, including interest on all outlay of Rs. 83,257, and I am very confident that a good 300-acre garden can as set out, be made for that sum. The rates assumed are so liberal that a fair margin is allowed for bad seed or any other misfortune.

(¹). The cost for seed, nurseries, and transplanting increases each year as the area, over which vacancies may exist, enlarges.

(²). The expenditure for planting the 100 acres each year includes cutting and clearing jungle, removing roots, digging, staking, pitting, and sowing the seed. In fact *all* expenditure including part of the pay of Manager and Establishment. The rate per acre *decreases* each year, because each year there is more expenditure of other kinds, which helps to pay for the Manager and Establishment.

(³). The reason why the rate for cultivation on the 100 acres planted each of the three first years increases each year, is given in the Table and remarks at pages 83 & 84.

COST OF MAKING A GARDEN.

Brought forward Rs.		83,257
4th year (1878).		
20 maunds seed, at Rs. 70 (⁴)	1,400	
Nurseries and labor transplanting (⁴)	500	
Repairs, buildings (⁵)	500	
Cultivating first 100 acres, at Rs. 80, second at Rs. 70, third at Rs. 60 per acre (⁶)	21,000	
		23,400
5th year (1879).		
10 maunds seed, at Rs. 70 (⁴)	700	
Nurseries and labor transplanting (⁴)	500	
Repairs, buildings (⁵)	500	
Cultivating first 100 acres, at Rs. 90, second at Rs. 80, third at Rs. 70 per acre (⁶)	24,000	
		25,700
6th year (1880).		
Nurseries and labor transplanting (⁴)	500	
Repairs, buildings (⁵)	500	
Cultivating first 100 acres, at Rs. 100, second at Rs. 90, third at Rs. 80 per acre (⁶)	27,000	
		28,000
7th year (1881).		
Nurseries and labor transplanting (⁴)	500	
Building a permanent Tea Factory and Tea Store and repairs to buildings (⁵)	12,500	
Cultivating first 100 acres, at Rs. 100, second at Rs. 100, third at Rs. 90 per acre (⁶)	29,000	
		42,000
8th year (1882).		
Nurseries and labor transplanting (⁴)	500	
New permanent houses for Manager and Assistant and repairs, buildings (⁵)	8,500	
Cultivating first, second, and third 100 acres, at Rs. 100 per acre (⁶)	30,000	
		39,000

(⁴) The seed to be bought is now less each year, as it is produced on the garden, and after the fifth year no more has to be purchased. From the fourth, and all subsequent years, nurseries for vacancies are calculated at Rs. 500 which is enough, as the garden has been previously yearly replenished. This expenditure will be continual as long as the garden lasts, for there will always be some vacancies to replace.

(⁵). Rupees 500 is a fair sum to estimate for ordinary annual repairs to buildings, and it will be required as long as the garden lasts. A temporary Factory was made in 1877 and a permanent building is now allowed for in 1881. Permanent Manager's and Assistant's houses are also allowed for in 1882. The garden can afford this now, for the profits are large. (See Table at page ——).

(⁶). For the rates assumed here see page 83.

9th year (1883), *and all years after.*

Nurseries, at Rs. 500 (⁴)		1,000
Repairs, buildings, at Rs. 500 (⁵)		
Cultivating the 3 acres, at Rs. 1 per acre (⁶) ... 30,000		
		31,000

Nothing is allowed for interest after the third year, for soon after that, *viz.* fifth year, the garden begins to give profits on the yearly operations.

All the above figures are carried out in the Table in the next Chapter page 171, and how large the profits on Tea may be will there be seen.

In none of the estimates of cost, up to this, is the expense of manufacturing the Tea included. It would have been very inconvenient to do so. The cost is so much per maund of Tea, and I prefer estimating the Tea at its market rate *minus* the cost of manufacture as shown at pages 162 and 70.

(⁴). The seed to be bought is now less each year, as it is produced on the garden, and after the fifth year no more has to be purchased. From the fourth, and all subsequent years, nurseries for vacancies are calculated at Rs. 500, which is enough, as the garden has been previously yearly replenished. This expenditure will be continual as long as the garden lasts, for there will always be some vacancies to replace.

(5). Rupees 500 is a fair sum to estimate for ordinary annual repairs to buildings, and it will be required as long as the garden lasts. A temporary Factory was made in 1877 and a permanent building is now allowed for in 1881. Permanent Manager's and Assistant's houses are also allowed for in 1882. The garden can afford this now, for the profits are large. (See Table at page 171).

(6). For the rates assumed here, see page 83.

CHAPTER XXX.

HOW MUCH PROFIT TEA CAN GIVE.

WE have already estimated the cost of making and cultivating a plantation of 300 acres. We must now ascertain how much Tea that area will give yearly.

It is a very wide question what produce an acre of Tea will give.

The following is an extract from the "Report of the Commissioners appointed to enquire into the state and prospects of Tea cultivation in Assam and Cachar," addressed to the Government of Bengal and dated March 1868 :—

"*Average produce per acre.*"

"The returns of actual produce of gardens in 1867 which we have obtained are so few in number that it is impossible to take any general average from them. The produce in these varies from three and-a-half maunds to one and-a-half maund per acres omitting the more recently formed gardens.

"From information received during our tour we have reason to believe that some gardens produce more than the highest rate per acre here mentioned; but, in the absence of returns of exact acreage and out-turn, we cannot notice these instances.

"Mr. Haworth, in his pamphlet already quoted, speaks of the produce of Cachar gardens as follows :—

"'I believe that three maunds per acre is fully one-third more than the present average yield of gardens in Cachar, after deducting the area of plant under yielding age.

"'There is no reason, that I am aware of, why the yield of Tea should not soon be raised to four maunds, and more gradually six maunds per acre, equal to twenty-four maunds of leaf per acre (less than one ton per acre for a green crop, which is still a very small one). Even now there are gardens in Cachar which give an average of from five to six maunds per acre this season. Some of these gardens have really no apparent advantage over their less fortunate neighbours, beyond that of a somewhat better system of cultivation and pruning; and these improvements even are to such a small degree ahead of the general practice, that I feel justified in saying I cannot place a limit on what the increased yield should be under a more rational system of cultivation, and the application of manures on a liberal scale, leaving out of consideration altogether what might reasonably be expected from a good system of drainage in addition.'

"Mr. James Stuart, Manager of the Bengal Tea Company's gardens in Cachar, has also given two maunds an acre as the general average of Cachar gardens for the past season, including young gardens of two, three, and four years old.

"We do not think it necessary to quote in detail the opinions of all the gentlemen examined by us on the subject of average produce per acre. A garden that can give four maunds per acre is undoubtedly a good one; and we have no doubt there are such or even better; but we do not think they are so common as to warrant our taking more than three maunds as a safe average."

Mr. A. C. Campbell, Extra Assistant Commissioner at Burpettah, in his "Notes on Tea cultivation in Assam," published in the Journal of the Agricultural and Horticultural Society of India, Part 3, Vol. 12, page 309, says:—"Good Tea land can be made to yield as high as seven maunds per poora." I forget exactly how much a poora is, but I believe it is near an acre.

In the Report to Government by the Commissioners, quoted above, at page 9, Mr. T. Burland, after estimating the cost of cultivation per acre per mensem at Rs. 9-10-2, adds:— "With the above expenditure per acre it is probable that much more than five maunds of Tea will be obtained from an acre of fair plant."*

---

* See my estimate for cultivation at page 83. I there estimate Rs. 100 per acre per annum from the sixth year, so that Mr Burland six years ago had come to the same opinion about high cultivation that I hold.

All these estimates, though but the last, are based on the cultivation of Tea as carried on hitherto with few exceptions, that is to say, on gardens covered with weeds for many months in the year, and to which no manure has ever been given. With such cultivation, particularly on gardens planted on slopes, I think myself that the yield will not exceed four maunds *at the outside*.

High cultivation and liberal manuring will, I believe, at least double the above, if the plants are of a high class. However here I give a Table on the subject which I have carefully framed.

Estimate of probable yield per acre on flat land, good soil, in a good Tea climate, and with hybrid plants, if really high cultivation and liberal manuring is carried out.

Year.	Supposed Year.			Estimated yield per acre in maunds.*
1st	1875	...	...	......
2nd	1876	...	...	......
3rd	1877	...	...	½
4th	1878	...	...	2
5th	1879†	...	...	4
6th	1880	...	...	5
7th	1881	...	...	6
8th	1882	...	...	7
9th	1883	...	...	7½
10th	1884‡	...	...	8

* Calculating Tea by maunds is convenient, inasmuch as pounds necessitate such lengthy figures for all calculations. The maund here employed is however quite an arbitrary measure. It is *not* the Indian maund, it equals and is represented exactly by 80 lbs. Any number of maunds multiplied by 80 will naturally give the lbs. of Tea.

† Up to this point, *viz.*, the fifth year inclusive, the figures given have been much more than realized and that on a garden with 15 per cent. vacancies. It has been though highly cultivated and liberally manured from the first.

‡ From the fifth to the tenth year is *understatement*, assumption, except that I know one garden which, to my certain knowledge, has given *more* than ten maunds an acre, and this in spite of about 15 per cent. vacancies. The garden is an old one, planted about 18 years ago. It is also a very small one. The soil is *very* poor, but the plants are of the highest class. It was much neglected till about eight years ago. From that time it has been highly cultivated in every way except in the point of irrigation, for it has not that advantage. It has been *most* liberally manured.

I do not think plants reach to perfect maturity under ten years. Certainly not with ordinary cultivation, perhaps with high cultivation, seven years would suffice.

That eight maunds per acre as estimated in the table just given *can* be realized, under the conditions stated, I have no doubt whatever, but I am equally certain that the size of some gardens in India must be much reduced if even five or six maunds are looked for. Not only must they be reduced in size, but they must be highly cultivated, must be manured, and no vacancies allowed. However, I have dwelt on all these points before, and need not repeat here, for unless the reader is convinced before this that a large area and low cultivation won't pay, it were waste to write more.

I now give a table showing the result for 12 years of a plantation such as I have advised.

PROFIT.   171

Table showing the estimated Results for 12 years of a 300-acre Plantation, in a good Tea climate, highly cultivated, and liberally manured.

1.	2.	3.	\multicolumn{5}{c}{YIELD IN MAUNDS OF 300 ACRES AND ITS VALUE.}					\multicolumn{3}{c}{YEARLY RESULTS.}			\multicolumn{4}{c}{FINAL RESULTS.}			REMARKS.
Year.	Yield per acre as per page 164.	Supposed year.	4. Rate yield of 100 acres planted in 1875.	5. Rate yield of 100 acres planted in 1876.	6. Rate yield of 100 acres planted in 1877.	7. Total yearly yield in maunds.	8. Value per maund.	9. Receipt from sale of Tea.	10. Total expenditure detailed at page 164.	11. Yearly Profit.	12. Yearly Loss.	13. Total receipts to end of each year.	14. Total expenditure to end of each year.	15. Balance to credit at end of each year.	16. Balance to debit at end of each year.	
	Mds.		Mds.	Mds.	Mds.	Mds.	Rs. 50 per maund, after cost of manufacture, packing, and transport are deducted, see pages 163 & 70.	Rs.	Rs.	Rs.	Rs.	Rs.	Rs.	Rs.	Rs.	
1	...	1875	...	...	...	...		...	22,600	...	22,600	...	22,600	...	22,600	It will be seen from this table as follows:—
2	...	1876	...	...	...	...		...	23,000	...	23,000	...	45,600	...	45,600	1. About Rs. 90,000 of capital is necessary to make a plantation as quick as this. If made more gradually, very much less would suffice.
3	½	1877	½	...	...	50		2,500	37,657	...	35,157	2,500	83,257	...	80,757	2. There is no yearly profit until the 5th year.
4	2½	1878	2	½	...	250		12,500	23,400	...	10,900	15,000	1,06,657	...	91,657	3. By the eighth year all the outlay is recovered.
5	4	1879	4	2	½	650		32,500	25,700	6,800	...	47,500	1,32,357	...	84,857	This table has been prepared with great care, and the authority for the figures assumed has been arrived at in previous parts. (See headings of Cols. for the pages and note at foot.
6	5	1880	5	4	2	1,100		55,000	28,000	27,000	...	1,02,500	1,60,357	...	57,857	
7	6	1881	6	5	4	1,500		75,000	42,000	33,000	...	1,77,500	2,02,357	...	24,857	
8	7	1882	7	6	5	1,800		90,000	39,000	51,000	...	2,67,500	2,41,357	26,143	...	I believe this table represents truly what Tea, with all the necessary advantages detailed at page 172, can do.
9	7½	1883	7½	7	6	2,050		1,02,500	31,000	71,500	...	3,70,000	2,72,357	97,643	...	
10	8	1884	8	7½	7	2,250		1,12,500	31,000	81,500	...	4,82,500	3,03,357	1,79,143	...	
11	8	1885	8	8	7½	2,350		1,17,500	31,000	86,500	...	6,00,000	3,34,357	2,65,643	...	
12	8	1886	8	8	8	2,400		1,20,000	31,000	89,000	...	7,20,000	3,65,357	3,54,643	...	

* With interest, see page 164.

	Cols.	Pages.
At the following pages will be found the calculations for the figures assumed	2	169
	8 }	163 & 70
	10	164

Y

The necessities for success in Tea are:—
1. A good climate.
2. A good site.
3. Perfect knowledge in Tea cultivation and Tea manufacture on the proprietor's part or that of his manager.
4. Seed from a high class of plants.
5. Local or cheap imported labor.
6. Facilities for manuring.
7. Cheap transport.

Do not dispense though with even *one* of the seven points named, for the truth is simply, that Tea will pay *very well* with all the above advantages, but will utterly fail without them.

Such is my advice to intending beginners. To those who have gardens, I say, reduce your areas till of the size you can really cultivate them highly, and procure manure at any cost.

I shall not have written in vain, and Tea enterprize in India will flourish, if the motto of planters in future be

"A full area, highly cultivated."

CHAPTER XXXI.

PAST, PRESENT, AND FUTURE OF INDIAN TEA.

A FEW words on the past, the present, and the future of Indian Tea will now conclude this Essay, and will, I hope, be acceptable to the reader.

The subject is one of growing importance, but being a new one, there are points connected with it on which the public are very ignorant, and should be enlightened.

To begin with, the following facts are not disputed by those who know anything of the subject:—

(1). Indian Teas have far more body, that is strength, than China Teas.

(2). Indian Teas consequently command a higher price, at the London sales, than China Teas.

(3). In spite of its higher price, it is far more economical than the China produce, as, generally speaking, one-third of the quantity suffices.

(4). There are lands enough in India to grow all the Tea required for England's use, and indeed for all her colonies.

If these *are* facts, and I confidently affirm they are so, how is it that the following holds in England?

(1). Indian Tea is not known to the public.

(2). Except in one or two shops in London and Glasgow, unknown to the mass of the people, not an ounce of pure Indian Tea can be bought in all England.

(3). That India is even a Tea-producing country is scarcely known in England.

I think I can explain some of these anomalies.

Tea is an acquired taste; by which I mean, not only that the adult who had never tasted Tea would not like it when first offered to him, but also that, with those who consume it regularly, any Tea that differs in flavor from what is habitually drunk is not relished.

It matters not whether it is intrinsically better or worse, enough that the flavor is different, for that reason it is not liked.

Indian Tea differs widely from China Tea, and for that reason is rarely appreciated by those accustomed to the latter. For a long time it appeared as if this difficulty would be a bar to the general introduction of Indian Teas in England, and so indeed it would have proved, had the short-sighted policy adopted at the commencement by one or two Indian Companies, that their Teas should be sold retail and pure, that is unmixed with China, been followed out. It did not avail to tell John Bull it was better Tea, that it was far stronger, that it was in no way adulterated; for he simply shook his head, the flavor was different to what use had made him familiar, and he would none of it.

But little by little, in spite of the above, it made its way. Grocers soon found that the worst *id est* the weakest class of China Teas received *body* and were made saleable by an addition of Indian Tea. It was not long after this that the trade discovered that pretty well *all* China Teas were improved, if proportions of Indian Teas were mixed with them. In short, the fact was recognized by Tea vendors that China Teas were weak, and much improved if mixed with Indian.

The public were thus *educated* to relish the superior flavor of Indian Tea, and did so, when the quantity mixed with the China was not so great as to make the new flavor too "prononcé." Little by little the custom of so mixing became very general, so much so that it may almost be said to-day that if Indian Teas cannot be purchased pure, no more can China. A mixture of China and Indian Tea, the latter small as compared with the former, is what is now generally used in Great Britain.

This is the case to-day. What will it be in the future?

As the English palate is educated to like the flavor of Indian Tea, more and more of it will be demanded in the mixture made up for the public, and though the day is distant, nay may never arrive on account of its greater cost; when it will be generally drunk pure, I do not myself doubt that the demand for it will go on steadily increasing for years to come, as it has for years passed.

It is an important query if, with a largely increased demand, the supply will be equal to it. Very far from all India has a good Tea climate, which is a peculiar one, and only exists in perfection in Assam, Cachar, Chittagong, and lands in Bengal close to the foot of the Himalayas.

But in these districts alone there are lands sufficient to supply nearly the whole world with Tea, so that it is not the lands which are wanting, though the Government prices for the lands are prohibitory and will check cultivation. But in Assam, Cachar, and the Terai below the Himalayas labor is very scarce, while in Chittagong the area fit for Tea is not large, so that I do not anticipate any very sudden increase of the cultivation, though year by year it *is* on the increase, and will so continue.

On the other hand, I do not, for the reasons stated, *viz.*, that Tea is an acquired taste, and thus a new kind is not at first palatable, anticipate any very sudden increase in the demand. If, however, I am wrong, and from a largely increased demand the prices of Indian Teas rise, I do not doubt that the cultivation will be quickly extended, and that after an interval of four years (it takes that time for the Tea plant to produce) the supply will be equal to the then wants of the English market.

The future of Indian Tea is, I think, a bright one, and I know nothing in which capital can be more profitably invested if the business is conducted with knowledge and experience, but to embark in it without these two requisites is ruination.

A few figures may be given here. The imports into Great Britain of Indian Teas have been yearly increasing, till in 1873 they amounted to 18,367,000 pounds, and judging from the estimate out here of the produce this year, *viz.*, 1874, the imports into Great Britain in 1874 will not be far short of twenty millions of pounds.

But as the annual consumption of Tea in the United Kingdom is not less than one hundred and thirty millions of pounds, India is still very far from supplying enough to give a mixture of ¾th China and ¼th Indian Tea.

The finest China Tea sells in London in bond at 2*s.* 4*d.* to 2*s.* 6*d.*, while the finest Indian in bond fetches 3*s.* to 3*s.* 6*d.*, and I believe the London Tea-dealers would hail with satisfaction the advent of much larger imports of the latter.

What then will be the future of Indian Tea? It is an important query. The industry is one which, if successful, might attain to wide limits, and help not a little to relieve the Indian State Exchequer, while it would afford occupation to many of a class of Englishmen who at present look about in vain for employment.

Tea speculation has passed through the first two preliminary phases to which most new ventures are liable. First we had the wild rush, the mad fever, when every man thought that to own a few Tea bushes was to realise wealth. In those days existing plantations were bought at eight and ten times their value; nominal areas of 500 acres were paid for, which, on subsequent measurement, proved to be under 100; new gardens were commenced on impossible sites and by men as managers who not only did not know a Tea-plant from a cabbage, but who were equally ignorant of the commonest rules of agriculture. Boards highly paid, with Secretaries still more liberally remunerated, were formed both in Calcutta and London to carry on the enterprise; and, in short, money was lavished in every conceivable way, while mismanagement ran rampant in each department. It is not strange that the whole thing collapsed, the wonder is it did not do so earlier.

The second stage was then entered upon. Numbers had been bitten, and the idea, once formed, grew apace that Tea could not pay at all. Every one wanted to sell, and down went all Tea shares to a figure which only increased the general panic. Many Companies, and not a few individuals, unable to carry on, had to wind up and sell their estates for whatever they would fetch. Gardens that had cost lakhs were sold for as many hundreds, and the very word 'Tea' stunk in the nostrils of the commercial public. A few of the best Companies held on, as also such individuals embarked in the speculation as could weather the storm; but some of the Companies were bowed down with heavy debts, and it has been, with many from that cause, a losing race ever since.

This great smash occurred seven years ago. I purpose, therefore, to examine into the future prospects of the industry, now that time has been given to test its vitality. Naturally the mistakes made at the first have not been repeated since, so the speculation has had more or less of a fair chance to show what it can do.

In the first place the Share List of Tea Companies in the public prints does not at all represent the true position of Tea property to-day. It only gives the dividends declared, and the value of the shares, in those few limited liability Companies which were able to weather the storm, but who, in common with all the others, were bowed down with debt, and are suffering up to the present time, both from that and the numberless mistakes made at the commencement of the enterprise. There are a few notable exceptions even among the Tea Companies. Some of these have done very well, pay large dividends and are quoted at a high premium, which shows that Tea can and will pay even with the disadvantages attached to limited liability Companies. I mean that in these latter work is always expensively done, and that much of the profits are swallowed up by Secretaries, Directors, &c., besides which generally from interested motives the Teas are sent home for sale which private planters know from experience is *not* the best plan.

But to return to the Share List. The very many gardens held by firms or private individuals are absent, and inasmuch as many of these were begun more lately, and consequently, the blunders made in other gardens were avoided, it is evident that *their* position, if it could be ascertained, would give the true picture needed.

There is one class of plantations which it would be by no means fair to include. I mean those gardens bought for a mere song during the panic. On many of these necessarily enormous profits have been made, but it proves nothing, inasmuch as the profits, to be legitimate profits for criticism, should on the debit side include the whole cost incurred in making the plantation. To form a fair appreciation of the profits Tea-planting can give, we must select gardens constructed after knowledge on the subject was attained, where good management, combined with economy in all details, has been carried out, and where the necessary natural conditions for success exists.

But first let me explain what I mean by the " necessary natural conditions for success." Manageable areas ; flat or nearly flat land for the garden ; a good class of indigenous and hybrid plants ; local labor or any how a good proportion of this ; facilities for manuring ; a good soil ; a good Tea climate ; and cheap means of transport constitute these, and where they exist I hold Tea *must*, and *does*, pay well. I don't believe in plantations of six and eight hundred acres; some of these pay, but they would pay much better if reduced in size. A garden of 300 acres, yielding even at the rate of four maunds an acre, will pay much better than another of 500 acres, yielding but two-and-a-half or three maunds. The reason is obvious, the larger produce is against a smaller expenditure. Were I to commence a Tea-plantation to-day, it should not exceed 300 acres in size. This passion for large areas is the rock on which, more than any other, Tea Companies have wrecked themselves; experience has already shown this, and will show it more, as time goes on.

Flat land for Tea gardens is a great desideratum. Steep lands are difficult to cultivate; the soil is continually washing away from the roots of the plants; it is impossible to manure them successfully, and the consequence of all this is that the Tea bushes do not thrive.

The China plant gives a small and inferior produce, the indigenous and hybrid kind a larger and very superior one; thus I think the latter one of the "necessary conditions for success." On the other points, with the exception of manuring, nothing need be said, inasmuch as their necessity is evident; but on the point of manure I must say a few words. The Tea-plant is being continually denuded of its leaves; nothing is returned to the soil; and consequently in process of time that soil is exhausted. It was held once that manure destroyed the flavor of Tea. This idea, at variance with all agriculture experience, is now completely exploded like many others received from the Chinamen who first came from the flowery land to teach the art of Tea-cultivation and Tea manufacture to the Indian public. Many of them had never perhaps seen a Tea bush, any how in many respects theirs was faulty teaching, and all experienced planters are convinced, and it is truth that more knowledge on Tea exists in Indian than China at the present time.

But to return to the subject of manure. generally allowed to be, a necessity to the successful maintenance of a plantation. Me tion are now largely adopted in Assam and results will be a yield per acre the most sa dreamt of. Chittagong, on this head, has great advantages; manure in any quantity can there be procured for a trifle: and the results have shown its great value.

We have scarcely yet entered on the third stage to which any new speculation, after the two first, (the wild venture, and the unreasoning panic have passed) tends, but as knowledge of the financial results of Tea-plantations in the hands of private firms and private individuals increases, that third stage will dawn if it has not done so already. It consists in a sober appreciation of the subject, opposed to both the extremely exulting and depressing views passed through, and when it arrives, the great and successful future of Indian Tea will be only a question of time.

FINIS.

www.ingramcontent.com/pod-product-compliance
Lightning Source LLC
Chambersburg PA
CBHW021733220426
43662CB00008B/832